Business Analytics

Series Editor

Klaus-Peter Schoeneberg, Fakultät Wirtschaft und Soziales, Rm 10., HS für
Angewandte Wissenschaften, Hamburg, Germany

The interdisciplinary approach of the series "Business Analytics" synergises the scientific disciplines of Ecommerce, business informatics, economics and mathematics. Based on the latest scientific methods, models and algorithms, the series integrates two juxtapositioned fields of research: scientific methodology and practically derived data in the discipline of strategic and operative business oriented questions. Within these disciplines, the series presents results of current research questions about data- and future-oriented themes as strategies, analytics and prediction. The findings of this series are highly relevant to all scientific fields in this area of research, academic institutions and companies that have an awareness of how data can drive business outcomes but want to better leverage data as a corporate asset for competitive advantage.

More information about this series at http://www.springer.com/series/15363

Matthieu Jaunatre

Renewable Hydrogen

Renewable Energy and Renewable Hydrogen APAC Markets Policies Analysis

Matthieu Jaunatre
Institute of Distance Learning
Beuth University of Applied Sciences Berlin
Berlin, Germany

This thesis has been written as part of the Master of Business Administration studies at the Beuth Hochschule Für Technik, Berlin, Germany.

ISSN 2570-1363 ISSN 2570-1371 (electronic)
Business Analytics
ISBN 978-3-658-32641-8 ISBN 978-3-658-32642-5 (eBook)
https://doi.org/10.1007/978-3-658-32642-5

Responsible Editor: Anna Pietras
This Springer Gabler imprint is published by the registered company Springer Fachmedien Wiesbaden GmbH part of Springer Nature.
The registered company address is: Abraham-Lincoln-Str. 46, 65189 Wiesbaden, Germany

Contents

Abbreviations

AEC	Alkaline water electrolysis
ARENA	Australian Renewable Energy Agency
BF-BOF	Blast Furnace – Basic Oxygen Furnace
CCS	Carbon Capture and Storage
CNREC	China National Renewable Energy Centre
CO_2	Carbon dioxide
CSIRO	Commonwealth Scientific and Industrial Research Organisation
DENA	German Energy Agency
DRI-EAF	Direct Reduction of Iron – Electric Arc Furnace
FCV	Fuel cell vehicle
GIZ	Deutsche Gesellschaft für Internationale Zusammenarbeit GmbH
GWEC	Global Wind Energy Council
H_2	Hydrogen
H_2O	Water
HHV	Higher heating value
HRS	Hydrogen refuelling station
IEA	International Energy Agency
IEAGHG	IEA Greenhouse Gas (R&D Programme)
IKI	International Climate Initiative
IPCC	International panel for the climate change
IPHE	International Partnership for Hydrogen and Fuel Cells in the Economy
IRENA	International Renewable Energy Agency
ITRI	Industrial Technology Research Institute (Taiwan)
LFCH	Levelized fixed cost of hydrogen
LRET	Large-scale Renewable Energy Target

METI	Ministry of Economy, Trade and Industry (Japan)
MOEA	Ministry of Economic Affairs (Taiwan)
NEDO	New Energy and Industrial Technology Development Organization (Japan)
NEV	New energy vehicle
O_2	Oxygen
PEMEC	Polymer electrolyte membranes
PPA	Power purchase agreement
ROC	Registered Organisations Commission (Australia)
RPS	Renewable Portfolio Standards
SDE++	Subsidie (Dutch) / Subsidy
SMR	Steam Methane Reformation
SOEC	Solid oxide
US EPA	United States Environmental Protection Agency
WTP	Willingness to pay

List of Figures

List of Tables

Introduction

1

1.1   The Definition of the Problem Statement

The author proposes to investigate the policy frameworks for renewable energy projects and renewable hydrogen production and how they contribute to the decarbonisation of industries using fossil fuel-based hydrogen as feedstock in targeted APAC markets.

There is a global trend to allocate renewable energy projects with the auction mechanism (IRENA 2019). Low auction price is an important criterion to win auctions. Moreover, the nature of the support provided to renewable energy project developers is changing. In Europe, generous feed-in tariff policies are gradually fading out and let renewable energy project developers exposed to the merchant risk (Heiligtag et al. 2018). In a context of fierce competition, renewable energy project developers start integrating new technologies to overcome issues such as curtailment or production intermittency and to maximise the revenues from their assets. One of those technologies is the production of hydrogen from electrolysis of water, otherwise called power-to-gas. Concurrently, 115 Mt yr^{-1} of hydrogen is used globally as feedstock in the petrochemical sector. Less than 0,7% of this hydrogen is produced from renewables or from fossil fuel plants equipped with CCUS (IEA 2019).

Form an academic perspective, grey and scientific literature covers in length the topic of hydrogen as the enabler of the energy transition, in the transportation sector, but also in the energy, heat and industry sectors. However, there is little research available on the underpinning policy frameworks and how those may influence the use renewable hydrogen for decarbonising industry sectors using fossil fuel-based hydrogen.

© The Author(s), under exclusive license to Springer Fachmedien Wiesbaden GmbH,
part of Springer Nature 2021
M. Jaunatre, *Renewable Hydrogen*, Business Analytics,
https://doi.org/10.1007/978-3-658-32642-5_1

From an industry perspective, the result of research can be used by renewable energy projects developers and hydrogen energy actors for further business development purpose.

Finally, from the author perspective, the choice for this thesis topic is motivated by a deep interest in renewable energy projects and the willingness to develop competence and knowledge in the field of renewable hydrogen.

1.2   The Objectives of the Research

The objective of the research work is to assess how well the current renewable policy frameworks support the replacement of the fossil-based hydrogen by the renewable hydrogen.

The first part of the research work is a review of the current policy frameworks supporting the development of renewable energy projects and renewable hydrogen production. The review covers the underpinning technical and economic aspects of producing and using fossil-based and renewable hydrogen.

Based on the review, the second part of the research consists on conducting a survey of the targeted countries regarding the status of the respective renewable energy and hydrogen policies.

Building-up on the results of the two first steps, the third and last part of the research is an assessment of the potential for decarbonisation of industry sectors using fossil fuel-based hydrogen as feedstock.

1.3   The Research Question

The research question is: What is the influence and impact of the renewable energy policies on the decarbonisation of industry sectors using fossil fuel-based hydrogen?

This primary question is answered in the section 7. Conclusions.

Secondary questions breakdown the primary question in logical steps that will be used to structure the research.

- What are the ambitions of the targeted countries for the development of their renewable energy and hydrogen production capacity?
  This secondary question is answered in the section 3.1. The renewable energy offer.

- What are the underpinning policies and instruments supporting the development of renewable energy and hydrogen?
  This secondary question is answered in the section 3.2. The policy instruments supporting the APAC renewable energy development and section 3.3. The review of APAC hydrogen policy landscape.
- What is the hydrogen demand of industry sectors using fossil fuel-based hydrogen as feedstock?
  This secondary question is answered in the section 4.2. The review of hydrogen production processes and in the section and in the section 5.3. Case studies for determining the potential of renewable hydrogen for decarbonising the industrial sectors using fossil fuel-based hydrogen within the APAC markets.
- What is the potential of renewable hydrogen for decarbonising the industry sectors using fossil fuel-based hydrogen as feedstock?
  This secondary question is answered in section 6. Results.

1.4   The Delineation of the Research Scope

For this research work, the following terminology will be used. Hydrogen produced from fossil fuels is named fossil fuel-based hydrogen. The hydrogen produced from renewable energy is named renewable hydrogen, where solar PV and wind power are the source of the renewable energy.

The research work focuses on of five countries of the APAC markets: China, Japan, South-Korea, Taiwan, and Australia.

The industry sectors using fossil fuel-based hydrogen considered in the scope of this research are the ammonia and methanol manufacturers, the oil refineries, and the steel manufacturers.

The Research Concept

2

2.1 The Purpose of Research

The proposed research has a combined descriptive and exploratory purpose. The descriptive research covers the review of the renewable energy policy instruments supporting the development of offshore wind, large onshore wind, and solar PV projects. Feed-in tariffs policy instrument and its derivatives has been a very efficient way to develop renewable energy. However, role model countries in the energy transition such as Germany are now reducing the extent of their support. The research work describes this phenomenon and its consequences for the renewable energy project developers. The descriptive research also provides an analysis of the renewable energy targets and policy instruments as well as the hydrogen economy of five case studies belonging to the APAC markets.

The results of the descriptive research are supplemented by an exploratory research consisting on assessing the opportunity for replacement of the fossil fuel-based hydrogen by the renewable hydrogen in the five targeted countries of the APAC markets.

The result of the combined descriptive and exploratory research is used to test the theory research claim: the renewable energy policy instruments have influence and impact on the decarbonisation of industry sectors using fossil fuel-based hydrogen as feedstock.

Electronic supplementary material The online version of this chapter (https://doi.org/10.1007/978-3-658-32642-5_2) contains supplementary material, which is available to authorized users.

2.2   The Research Philosophy

The four research paradigms defined by Burrell and Morgan (Burrell et al. 2019) aim at categorizing the philosophy of a research and offering some foundations for its design. For Burrell and Morgan, the four research paradigms are mutually exclusive. However, mutual exclusivity is difficult to achieve in practice and multi-paradigmatic research design can be indicated and even required (Taylor, Taylor, and Luitel, 2012).

Therefore, the research work refers to the interpretive paradigm and to the functionalist paradigm. The interpretive paradigm is characterized by a need to understand a phenomenon occurring in the complex but orderly social world created by human beings. In this paradigm, the researcher is usually part of the research framework and is not an objective observer of the phenomenon being investigated.

The first part of the research investigates the status of renewable energy policy instruments. In this context, it is important to understand how the renewable energy policy instruments impact the renewable energy developers when bidding for an auction and how this leads to the horizontal integration of renewable hydrogen. By developing this line of thought, the researcher guides the research process which is a trait of an interpretive research. Investigating the phenomenon described above also requires interacting with experts in the domain which will infer some new perceptions in the research process.

Interpretive research is well suited for multifaceted social processes such as the consequences of shifting from one policy instrument to another. It is also appropriate for studying context specific and unique processes which can be further developed in exploratory research (A. Bhattacherjee 2012).

The functionalist paradigm, or positivism, is also relevant in scope of the proposed research. Data is required to assess the potential of an organisation to decarbonise its activities by replacing fossil-based hydrogen by renewable hydrogen. The use of economic and technical data does not fit with the interpretive research paradigm. For this aspect of the research work, the researcher is independent and impartial from the economic and logistics rationales of the hydrogen.

2.3   The Research Approach

The research work follows an inductive approach. The phenomenon observed is the potential influence and impact of renewable energy policy instruments on the decarbonisation of industry sectors using fossil fuel-based hydrogen energy.

The descriptive research focuses on analysing the influence and impact of the renewable energy policy instruments on the development of large renewable energy projects and the integration of renewable hydrogen production in their value chain. This is done following an interpretive research philosophy. Literature review and interviews of renewable energy policy experts as well as representatives from renewable energy project developers will be used to support the analysis.

The descriptive research focuses on a cross-section of five case studies identified in the APAC markets. The cross-section, performed as a survey, reviews the current and planned renewable energy targets. It reviews the policy instruments currently in use and if any development is expected. Finally, the cross-section addresses the hydrogen off-take of the case studies. The survey conducted on the case studies is done following a positivist research philosophy. The research is based on data analysis from the renewable energy and hydrogen laws and strategies and related publications. Publications from energy agencies are also analysed. Hydrogen off-take data is taken from official publications from the case studies' relevant governmental authorities and relevant industry associations.

The exploratory research aims at identifying relationships out of the descriptive research results to test the theory research claim.

The research approach is illustrated in the Thesis Conceptual Framework available at the appendix 1.

2.4   The Research Instruments

The research instruments used in scope of the proposed research are the documentation review, the expert interview, the survey of case study.

The documentation review and the expert interview research instruments are used with an interpretive research philosophy to perform a descriptive research of the renewable energy and hydrogen policies.

The survey of the case studies is used with a positivist research philosophy to assess the potential of decarbonisation of fossil fuel-based industry sectors by using renewable hydrogen.

2.4.1 Documentation Review

The documentation review consists on getting insight, through reading relevant publications, on the phenomenon of renewable energy policy instruments status. The documentation review shall answer the secondary questions 1 and 2.

2.4.2 Expert Interview

The expert interviews are done by using a semi-structured questionnaire to bring further depth and perspective to the documentation review as well as corroborating some of its findings. For that purpose, the expert interviews are conducted after the documentation review is completed. Three types of experts are considered for the interviews. The renewable energy policy instrument experts provide information about the policy instruments. The renewable energy project developers provide information regarding the impacts of the policy instruments on the development of renewable energy and the production of renewable hydrogen. Finally, off-takers provide information regarding the drivers for substituting fossil fuel-based hydrogen by renewable hydrogen. The findings of the semi-structured interview are identified as described in the appendix 17—Interviews analysis framework.

The expert interviews will further substantiate the secondary question 1.

Face to face interviews are preferred to ensure a good cooperation with the interviewees. When personal interviews are not possible, mainly for limiting costs linked to travelling, interviews are conducted virtually, by videoconference or by phone.

2.4.3 The Survey of the Case Studies

Case studies can be used for theory building or theory testing (A. Bhattacherjee 2012). In the case of the present research, the case study is used for theory testing. A case study is a research strategy, which focuses on understanding the dynamics within single settings (Eisenhardt 1989). Case studies typically combine data collection methods such as document review, interviews, questionnaires, and observations. Finally, evidence may be qualitative, quantitative, or both. The proposed research uses the survey as instrument to frame the collection of data for the renewable energy targets and policies as well as the hydrogen off-take of the case studies (five countries of the APAC markets). The data collected consists of quantitative evidences.

The case studies of the proposed research can be qualified as exploratory case studies. The case studies are undertaken before implementing further large-scale investigations when there is considerable uncertainty. It is not sure that the case studies offer renewable energy legal and economic framework for renewable energy project developers, neither that the case studies have a developed hydrogen economy.

Case studies are well suited research instrument for the descriptive part of the research conducted with a positivist research philosophy.

The case studies will answer the secondary questions 3 and 4.

2.4.4 Target Populations/Sampling

Nonprobability samples are used for the expert interviews and the case studies.

2.4.5 Samples of Expert Interviews

The author has the objective to interview five convenience samples of policy experts, five convenience samples of renewable energy project developers and five convenience samples of hydrogen off-takers. Following the first interviews, possible snowball samples will be identified.

For generalization purpose, samples should cover the main regions where large renewable energy projects are currently developed. For example, experts interviewed could be from different regions such as Europe, North America, or APAC, enabling the analysis of the phenomenon at a global scale.

2.4.6 Samples of Case Studies

The author plans to conduct the survey on the five targeted countries of the APAC markets. Those are judgement samples. The countries have been selected based on their ambitious renewable energy plans and their developed economy favourable to the consumption of fossil fuel-based hydrogen.

2.4.7   Data Collection and Analysis

Data collected through documentation review and expert interviews will be subject to qualitative analysis to answer the secondary questions 1 and 2.

The documentation review will be done by desktop review of publicly available research work, articles, and journals about the phenomenon.

The expert interviews will be conducted through a semi-structured questionnaire, enabling some room for improvisation while covering the key areas of research.

Data collected for the case studies are subject to quantitative analysis to answer the secondary questions 3 and 4. The data collection is done by a desktop review of relevant publications. To enable consistency of the quantitative analysis across the case studies, a survey is designed to capture objective data, for example installed capacity in GW, target capacity in GW, quantity of hydrogen in tons…

The analysis is done by methodological triangulation of the data. The data captured during the survey of the case studies will be used to validate the findings from the documentation review and the expert interviews.

2.5   The Renewable Hydrogen Framework

To provide a structure for conducting the research (Lambert 2008; N, Dr. Elango-van 2015), the author has defined the *renewable hydrogen framework*. It provides a structure to conceptualise the different building blocks of the research topic (Figure 2.1).

The "renewable energy offer" box represents the renewable electricity generation capacity in the countries considered in the research work. The overview of the renewable electricity capacity as well as the drivers and the challenges for its development is essential to assess the ability to produce renewable hydrogen from electrolysis of water.

The "hydrogen demand" box represents the hydrogen demand or off-take from carbon intensive industry sectors using large quantity of fossil fuel-based hydrogen as feedstock. The research work focuses on the hydrogen demand for ammonia, methanol, oil refining and steel production industry sectors.

The "potential for decarbonisation" box connects the renewable energy offer, enabling the production of renewable hydrogen, with the hydrogen demand. This part of the framework aims at understanding the ability of the renewable energy sector to decarbonise carbon intensive industry sectors using fossil fuel-based hydrogen as feedstock.

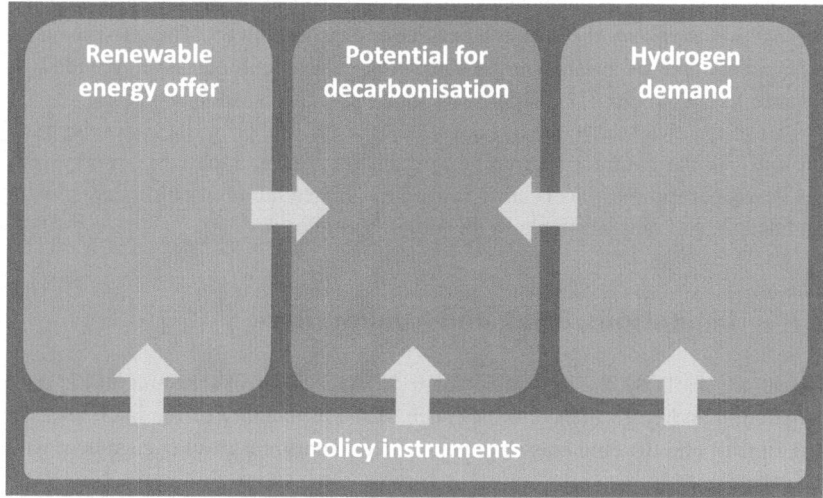

Figure 2.1   The renewable hydrogen framework

The underpinning "policy instruments" box aims at understanding the nature of the relevant policy supports and mechanisms developed by the governments of the targeted APAC countries. Eventually, the research aims at determining if the policy mechanisms supporting the renewable energy and the renewable hydrogen production are developed in isolation or if they are integrated to support the decarbonisation of carbon intensive industry sectors.

The three research instruments discussed in section 2.4 are utilized to explore the *renewable hydrogen framework* in the five targeted countries:

(1) The documentation review of the relevant grey and scientific literature,
(2) The expert interview in a form a semi-structured interview,
(3) The survey of the APAC countries targeted for the current research work.

2.6   Access, Privacy, Confidentiality, and Ethics

The author is committed to carry out the proposed research work in full compliance with the ethical standards set by the Beuth Hochschule University of Applied Sciences.

When contacting experts, the author ensures full transparency regarding the purpose and the use of the data collected during the interviews. The questionnaire, background and intentions of the author is clearly outlined and provided in advance to the experts. The experts are interviewed on a voluntary basis.

Part of the quantitative analysis may be used for further business development purpose. As the author is currently working for a large renewable energy project developer, conclusions bearing sensitive commercial information may remain confidential and only available to the author's employer.

2.7   Limitations, Risks, and Assumptions

The author shall take the limitations related to interviews into account during their preparation and performance. Issues such as the artificiality of the interview, the lack of trust and the elite bias shall be avoided to ensure fruitful engagement with the experts.

The author is currently employed by a large renewable energy project developer. This may have an impact of the sample selection as there are little chance for the researcher to interview renewable energy project developer representatives outside his own organisation.

Practical Literature Review for Establishing the Current Renewable Hydrogen Framework

3

The *renewable hydrogen framework* described in the section 2.5 is used by the author to structure the practical literature review. The practical literature or grey literature consists on non-scientific publications and articles. The author performed the research based on the two following criteria:

- The publications are issued from the year 2017 on.
- The key words corresponding to the *renewable hydrogen framework* elements were used to search the publications and articles searched on a web browser (Google®).

The practical literature has focused on the renewable energy offer and the policy instruments parts of the renewable hydrogen framework, Figure 3.1.

3.1 The Renewable Energy Offer

The practical literature review of the renewable energy offer consists on the review of the renewable energy political framework defined by the five targeted APAC countries.

3.1.1 Australia

Although coal and gas are the main sources of electricity production, the Australian Government is committed to reduce the GHG emissions and to encourage the development of sustainable and renewable sources of energy (About the Renewable Energy Target n. d.). The Renewable Energy Target is a scheme that covered

M. Jaunatre, *Renewable Hydrogen*, Business Analytics,
https://doi.org/10.1007/978-3-658-32642-5_3

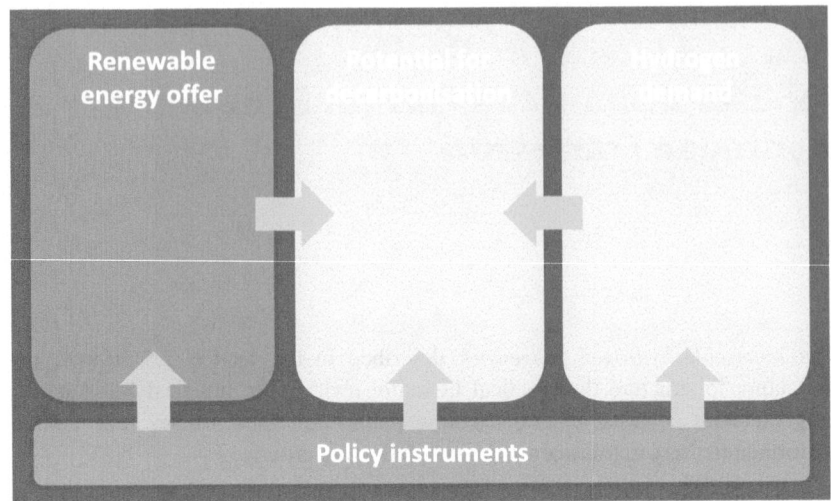

Figure 3.1 Focus areas for the practical literature review

large-scale power stations but also owners of small-scale systems. The Large-scale Renewable Energy Target (LRET) is set to reach 33.000 gigawatt hours of additional generation by 2020. The 30.000-gigawatt hours target remains constant from 2021 to 2030 under the Renewable Energy Act 2000 (Federal Register of Legislation—Australian Government n. d.). The Renewable Energy Target scheme is a nationwide programme completed by the Australian Federal States own renewable energy targets.

New South Wales

New South Wales has an energy security issue. It is the most coal dependent state of Australia for its electricity generation. New South Wales imports electricity from Queensland and Victoria. The strategy document suggests the creation of 3.000 megawatts Renewable Energy Zone opened to renewable energy project developers (NSW Government 2019).

Victoria

The State of Victoria has set the renewable energy targets to 20% of the energy mix by 2020 and to 40% by 2025. These targets have been amended to reach 50% by 2030 (Renewable Energy Action Plan 2017). The renewable energy generation capacity to reach those targets is estimated to be of 10 gigawatts from 2020 to 2030, mostly from wind (Victorian Renewable Energy Transition Economic Impacts Modelling 2019).

Northern Territory

In 2017, renewable energy accounted for 4% of the Northern Territory energy mix. The Northern Territory government has set a target to reach 50% of renewables by 2030. This would equate to approximately 900 gigawatts hour of renewable energy production per annum. This requirement would be covered by 450 megawatts of solar PV (Langworthy et al. 2017). The Northern Territory has the potential for the development of 10 gigawatts of renewable energy production capacity by 2030, corresponding to 20 times the current targets (Beyond Zero Emissions 2019).

South Australia

There is no official target for renewable energy for South Australia despite its visionary renewable hydrogen ambitions. The South Australian state had a share of 43.4% of renewable energy in the energy production mix in 2017 and it is on track for a 73% share by 2020 (Climate Council of Australia 2018).

Western Australia

The government of Western Australia has not defined renewable energy targets. The share of renewable energy in the energy production mix was of 8.2% at the end of 2018 (M. Maisch 2018). To reduce the GHG emissions in accordance with the 1.5°C climate change target, the Western Australia state should reach carbon neutrality by 2050. This would entail that the renewable energy share in the energy mix should be of 30% by 2025, 90% by 2030 and 100% by the early 2030 s, including phasing out coal before 2030 and gas shortly afterwards (Climate Analytics 2019). A 15 gigawatts wind and solar project is considered at Pilbara to produce renewable hydrogen via water electrolysis (G. Parkinson 2019).

Tasmania

The Tasmanian State has a renewable energy target set to 100% by 2022 (Climate Council of Australia 2018). The Tasmanian State already has a 90% share of renewable energy in its energy production mix. The renewable energy is produced mainly from hydropower with a 90% share followed by wind power with a 10% share. The Tasmanian government would like to increase it renewable energy production by a 1.000 gigawatts hour to become a next exporter of electricity (Department of Premier and Cabinet n.d.).

Queensland

The Queensland State has a renewable energy targets set to 50% by 2030. The state has the largest number of projects under construction (Climate Council of Australia 2018).

Australian Capital Territory
The renewable energy targets of the Australian Capital Territory are set to 100%
by 2020 and the State is on target to reach this objective (Climate Council of
Australia 2018).

3.1.2 China

The renewable energy targets for China are defined in three successive Five-Year
Plans. The 14th Five-Year Plan (2020–2025) aims at scaling-up the renewable
energy industry with a share of renewable energy in the energy mix set to 19%.
The 14th Five-Year Plan also defines annual additions of wind power generation
by 53 gigawatts and solar PV generation by 58 gigawatts. The 15th Five-Year
Plan (2025–2030) suggests an annual additional wind power generation capacity
of 127 gigawatts and 116 gigawatts of solar PV. Finally, the 16th Five-Year Plan
(2030–2035) defines an annual addition of about 150 gigawatts of wind and solar
energy. China has defined critical targets for renewable energy such as by 2025,
the installed wind power capacity should reach 500 gigawatts, equivalent to a
1350 terawatts hour of electricity generation on a yearly basis. The solar PV
installed capacity should reach 530 gigawatts, producing around 690 terawatts
hour of electricity generation (China renewable Energy Outlook 2019).

3.1.3 Japan

The Japanese renewable energy targets are set to 22–24% of the energy mix by
2030. In 2017, the share of renewable energy in the energy mix was 17.4%. This
included the small and large hydro energy production, accounting for about 50%
of the total renewable energy production.

The cumulative capacity of wind power was of 5,4 gigawatts in 2017. The
Japan Wind Power Association has defined a wind road map with an onshore
wind capacity and an offshore wind capacity respectively of 10 gigawatts and 0,7
gigawatt by 2020, 27 gigawatts and 10 gigawatts by 2030 and 38 gigawatts and
37 gigawatts by 2050.

For solar PV, the installed capacity was of 42 gigawatts at end of 2016.
The Japan Photovoltaic Energy Association expects an installed capacity of 66
gigawatts by 2020 and 100 gigawatts by 2030 (Institute for Sustainable Energy
Policies 2019).

3.1.4   South Korea

In 2017, the solar PV capacity was of 5.5 gigawatts and the wind capacity was of 1.1 gigawatts. The South-Korean government is committed to increase the share of renewable energy in the generation mix from 7% to 20% by 2030. This corresponds to an additional 30,8 gigawatts of solar PV and 16.5 gigawatts of wind capacity. Large-scale projects are planned to account for 28.8 gigawatts of this additional capacity (Lee 2019).

3.1.5   Taiwan

The Taiwanese government has defined three key objectives by 2025: (1) reduction of the coal-fired and oil-fired energy production, (2) nuclear-free energy production and (3) increasing the renewable share of energy production from 4.8% in 2016 to 20% by 2025.

The Four-Year Plan of Promotion for Wind Power sets targets for both onshore and offshore wind with respectively 814 megawatts and 520 megawatts by 2020, corresponding to 1,9 terawatts hour of electricity generation each (Bureau of Energy, Ministry of Economic Affairs 2009). Further development brings the targets for onshore and offshore wind power capacity respectively to 1.2 gigawatts and 3 gigawatts by 2025, corresponding to 2,9 terawatts hour and 11,1 terawatts hour of electricity generation.

Regarding solar PV, a target of 6,5 gigawatts of power capacity is set for 2020 corresponding to 8,1 terawatts hour of electricity generation. In 2025, the solar PV capacity is expected to be of 20 gigawatts for a 25 terawatts hour of electricity generation (R. Wang 2017).

In November 2019, the Taiwanese government took the decision to increase the offshore wind capacity targets. 10 gigawatts of additional offshore wind are now planned between 2026 and 2036, bringing the capacity to 15.5 gigawatts in 2035, up from the 5.5 gigawatts in development by 2025 (Qiao 2019).

3.2   The Policy Instruments Supporting the APAC Renewable Energy Development

The author reviewed first some general publications concerning auctions and renewable energy policy instruments. The practical literature review continued with the applicable support mechanisms available in the relevant APAC markets for new large-scale renewable energy projects.

3.2.1 The Auctions and Renewable Energy Policy Instruments

Auctions are becoming the most utilized support policy for the allocation of large-scale renewable energy projects. An auctioneer determines a certain amount of renewable capacity or energy to be built in a specific period. The auction will determine which renewable energy developer will be able to deliver the project on time for the lowest amount of support. The renewable energy developer can be awarded the project based on the lowest price. This is a "pure-price" based auction. The allocation can also be based on several criteria, also called a "beauty contest". An auction is not a support mechanism as such. It is only one element of a wider general support system. Feed-in tariffs can still be used as the main support instrument while the auction is used to allocate the project to the most cost-effective project developer (Alterman 2018).

The auction system provides an advantage to both the auctioneers and the renewable energy project developers. In case of product-based strategy, the auction leaves the opportunity for the renewable energy project developers to propose flexible resources to comply with the electricity market product requirements (i.e. storage system…) (Heiligtag et al. 2018).

In Europe, the renewable energy market has seen the energy policy instruments transitioning from fixed feed-in tariffs (Germany 1991–2016) to competitive auctions with subsidies (UK from 2015) and very recently to subsidy-free tenders (Netherlands from 2017). The auction systems with subsidy-free and market-based premium will become the standard by 2030. The market is currently adapting to this new policy landscape where the remuneration value is linked to the competition level (Enel 2016). The shift in policy instruments create new risks for the renewable energy projects developers. The auction system increases the allocation risk. It also exposes the developers and investors to the volatility of the electricity market price (AURES Consortium 2019). Consequently, renewable energy project developers are now exposed to merchant risk as their revenues is linked to the wholesale electricity market price. Renewable energy project developers are pushed to increase the quality of their revenue design.

A trend towards corporate power purchase agreements (PPAs) can be observed in some markets, in Europe, US and in the APAC region (CNREC 2019).

In this context, hydrogen production from renewable electricity becomes another revenue stream for large-scale renewable energy project developers. The

Danish company Ørsted decided to include the production of renewable hydrogen in the tender for the Holland Coast South 3&4 offshore wind farm (Orsted 2019).

3.2.2   The Renewable Policy Instruments of Australia

The main policy instrument to support the development of renewable energy is a renewable electricity generation certificates mechanism (About the Renewable Energy Target n.d.). The Renewable Energy Target is an Australian Government scheme to encourage additional generation of electricity from sustainable and renewable sources. There are two mechanisms in place, one for large-scale renewable energy projects and one for small-scale renewable energy projects. In May 2018, The LRET incentivises the development of renewable energy projects by the creation and sale of large-scale generation certificates (LGCs). The number of LGCs to be issued on a yearly basis are defined through a renewable power percentage system operating under the authority of the Clean Energy Regulator (19,31% in 2020, equivalent to 33,7 million LGCs).

A renewable energy project developer shall get the power station accredited under the RET. Once accredited, the renewable energy power station is entitled to create LGCs and sell them.

In addition to the LGCs mechanism, renewable energy projects secure revenue streams through PPA (Briggs 2019).

3.2.3   The Renewable Policy Instruments of China

The recommendations of the 14th Five-Year Plan stipulate that the supporting renewable energy policies shall be ensured after a transition from the feed in tariff mechanism to market prices. Although the new supporting policy mechanisms have not yet been defined, the report mentions the reduction of the regulatory costs imposed on renewable energy companies. The recommendations of the 14th Five-Year Plan focuses on the electrification of the industry to reduce the coal consumption (China renewable Energy Outlook 2019).

3.2.4   The Renewable Policy Instruments of Japan

The feed in tariff policy instrument is the support mechanism for both solar PV and wind power generation as described in the Strategic Energy Plan issued in

2018 by the Agency for Natural Resources and Energy (Agency for Natural Resources and Energy n.d.).

3.2.5   The Renewable Policy Instruments of South Korea

The Korean Energy Agency supports the development of renewable energy projects through two main mechanisms: the feed in tariffs and the RPS (KEA—KOREA ENERGY AGENCY n. d.). The targets for the RPS, defined under the Enforcement Ordinance of the Act on the Promotion of the Deployment, Use and Diffusion of New and Renewable Energy, are gradually increasing in order to reach 10% of the total power generation of electricity utilities by 2022.

3.2.6   The Renewable Policy Instruments of Taiwan

While the allocation of renewable energy projects is done through an auction system, the Taiwanese government has established a feed-in tariff system with a price ceiling (Couture et al. 2015). The feed-in tariffs are defined for onshore and offshore wind as well as for solar PV projects (Ministry of Economic Affairs 2019).

3.3   The Review of APAC Hydrogen Policy Landscape

The research concerning the hydrogen policy landscape focused on the targeted APAC countries strategies or roadmap publications. Some supporting reports and papers referenced in the country specific strategy or roadmap were also reviewed.

3.3.1   Australia

At a Federal level, Australia has the ambition to be the frontrunner for the production and exportation of hydrogen. The Australian National Hydrogen Strategy document provides a review of the ongoing and planned actions undertaken at a federal level and by each of the Australian states (Commonwealth of Australia 2019). The document does not use the term renewable hydrogen but clean hydrogen. The hydrogen production from water electrolysis is one of the paths considered in the document. However, the strategy refers also to production of hydrogen from coal gasification and SMR with CCS enabling carbon emissions abatement. The strategy promotes an adaptative pathway consisting on demonstrating and developing the technologies and the supply chains until 2025. Between 2015 and 2019, Australia has invested over AUSD 146 million in R&D, feasibility, demonstration, and pilot projects. From 2025 on and once solid foundations are established, the strategy entails a large-scale production, supply chains and market development, for domestic purpose, but especially for exportation purpose. The strategy relies on a possible growth of global hydrogen demand from 72 million tons in 2020 to 77 million tons in 2030 and reaching possibly 300 million tons by 2050. While Australia currently produces 0,5 million tons of hydrogen per year for domestic use only, 0,026 to 0,345 million tons could be exported by 2025, 0,242 to 1,088 million tons by 2030 and 0,621 to 3,180 million tons by 2040. The targeted importing countries would be Japan, Korea, and China. Japan would account for more than 60% of the Australian hydrogen exportation (Acil Allen 2018). If this plan is realized, up to 7.600 jobs could be created as well as AUSD 11 billion in GDP by 2050. Australia considers developing so-called "hydrogen hubs" and sector coupling as key contributors for the realisation of their strategy. The report identifies the public awareness as a key success factor.

When considering hydrogen produced from renewable energy, 11% of Australia territory could be used for solar PV, wind, and hydropower production (872.000 km^2). However, electrolysis requires water and therefore renewable hydrogen production would be better suited on coastal areas while using desalinated water. This would require 3% of the Australian territory surface and could cover the entire global demand in 2050 as estimated by the Hydrogen Council. The report suggests that between 2018 and 2020, Australia has developed its renewable energy production capacity four to five times the rate observed in Europe, Japan, US, or China. 10 gigawatts of wind power capacity should be installed by end of 2020 and close to 30 gigawatts of solar PV.

Australia is a net exporter of coal and natural gas. The strategy emphasises the great potential of hydrogen production through coal gasification and SMR associated with CCS. The document entails that a 90% CO_2 sequestration rate in depleted oil and gas fields or in salt caverns is technically feasible.

The demand for water maybe a challenge in case of large-scale hydrogen production. Coal gasification and water electrolysis both require 9 kg of water per kg of hydrogen produced while SMR requires 4,5 kg of water per kg of hydrogen produced.

In an evaluation of the economic viability of a sustainable hydrogen supply chain, the engineering consultancy firm Jacobs has compared three business models that could be implemented in the Australian context (Jacobs 2019). The first model consists on building dedicated renewable energy plants for the sole purpose of producing hydrogen. The second model considers the purchase of renewable electricity through PPA. The third and last model consists on using curtailed production from renewable energy plants. The report indicates that the last model is the least economically viable. Purchasing renewable energy through PPA offers the greatest economic alternative, allowing flexibility and scaling-up in case the demand for renewable hydrogen increases. The paper mentions that sustainable water inputs required in the electrolysis process can be a challenge in some parts of Australia.

Most of the Australian States have defined their respective hydrogen roadmaps or strategy documents (Energy Magazine 2020).

Queensland

The Queensland State defines its competitive position on the hydrogen market in the "Queensland Hydrogen Industry Strategy 2019–2024" (Queensland Government 2019). While a definition of the renewable hydrogen as hydrogen produced without fossil fuels is provided, the short-term and mid-term hydrogen production technologies considered are the natural gas reformation and the coal gasification. Water electrolysis along with other non-fossil fuel-based production technologies (i.e. high temperature pyrolysis) are considered but in smaller proportion and in longer term (TIQ Australia 2019). The strategy document supports the development of a hydrogen-based fuels green certification system. The document highlights the 300 days of sunshine of Queensland. 1.380 megawatts of solar PV are already installed, and a further 19 gigawatts of solar PV projects could possibly be developed. The development of a hydrogen economy would enable the creation of about 2.700 jobs by 2030 ad over 7.000 jobs by 2040. There is no target defined for either production of renewable energy or production of renewable hydrogen. The strategy suggests a set of enabling actions that are structured around five focus areas: (1) supporting innovation, (2) facilitating private sector investment, (3) ensuring an effective policy framework for sustainable development, (4) building community awareness and confidence and (5) facilitating skills development for new technologies. The Queensland Hydrogen Strategy includes an AUSD15 million Hydrogen Industry Development Fund.

New South Wales

The New South Wales government has announced a target of a 10% share of hydrogen in its gas network by 2030 (Energy Magazine 2020).

Victoria

The Victoria state hosts the HESC Project consisting on the production of hydrogen from coal gasification with CCS (Hydrogen Energy Supply Chain 2020).

Regarding renewable hydrogen, the Victoria state is currently in a public consultation phase to define the Victorian Green Hydrogen Industry Development Plan (The Victorian Hydrogen Investment Program : Engage Victoria n. d.). In order to provide information for the public, the Victoria State government has issued a "Green Hydrogen Discussion Paper" (Victoria State Government 2019). The document discusses the advantages offered by the Victorian state to produce renewable hydrogen. The renewable energy targets are mentioned, and the document entails that some new renewable energy production sites have been identified. The discussion paper refers to the HESC Project and to the experience gained in developing the hydrogen supply chain. The main applications identified are the decarbonisation of industrial feedstocks and the transport system. No hydrogen production targets are provided.

Northern Territory

The Northern Territory has no hydrogen strategy. However, the renewable hydrogen could be one of the opportunities for economic development of the state. By 2030, the Northern Territory could develop 4.87 gigawatts of renewable hydrogen energy production capacity accounting for 60% of the Australian hydrogen for export (Beyond Zero Emissions 2019).

Western Australia

The Western Australia State has identified the opportunity of the global growing demand for hydrogen (Department of Primary Industries and Regional Development 2019). The State benefits from numerous advantages for the successful development of the hydrogen industry. Western Australia has high-intensity renewable energy resources. There are 2.5 million km^2 of low intensity land available for the development of large-scale renewable energy generation. Western Australia benefits from a well-developed industrial and export infrastructure that could be used for the hydrogen industry. Its geographical position gives Western Australia an advantage due to its proximity to Asia and to the main potential importing countries such as Korea and Japan. The main strategic focus areas are the export, the remote applications such as mining, the blending in the natural gas network (10% of renewable hydrogen blend by 2040), and the transport sector. The Western Australian State has allocated an AUSD 10 million Renewable Hydrogen Fund. Beside the natural gas network blending target, the strategy document does not provide renewable energy or renewable hydrogen targets.

South Australia

Through the South Australia's Hydrogen Action Plan, the Government of South Australia has positioned itself as a pioneer for the development of the renewable hydrogen production. With an abundant renewable energy resource, the State has built an ambitious renewable energy projects pipeline. Over AUSD 7 billion have been invested in the renewable energy and AUSD 20 billion are in the pipeline. 14 gigawatts of renewable energy generation and storage projects are being considered (Government of South Australia 2019). South Australia has committed AUSD 17 million in grants and over AUSD 25 million in loans for renewable hydrogen projects. South Australia has directly contributed to the development of renewable hydrogen projects:

- Port Lincoln ammonia supply chain demonstrator (AUSD 4,7 million in grants and AUSD 117,5 million in loan) to produce a yearly 18.000 tons of ammonia from wind and solar PV power.
- Hydrogen Superhub at Crystal Brook energy park with first a feasibility study (AUSD 1 million grant) for a 50 MW renewable energy park producing a daily 25 tons of hydrogen from solar and wind power. A further AUSD 4 million in grants and AUSD 20 million in loans would be allocated by South Australia in case the project would be realized.
- Hydrogen Park South Australia with a 1.25 MW PEMEC electrolyser to produce renewable hydrogen for a 5% gas network blending supplying 700 homes (AUSD 4,9 million grant). The objective is that the gas network of South Australia should reach a net-zero carbon emission by 2050.

South Australia has the ambition to integrate hydrogen into its energy system while becoming a net 100% renewable energy producer in the 2030s. Domestic applications are expected to be mature by the mid-2020 s while large scale export projects are expected to be established by the early 2030s.

Beside the concrete examples mentioned above, the strategy does not provide renewable hydrogen targets nor budgets for further grants and loans available to developers.

Tasmania

The draft of the Tasmanian Renewable Hydrogen Action Plan lays the renewable hydrogen goals for the State (State of Tasmania 2019). By 2022, renewable hydrogen should be produced in Tasmania for domestic use while export projects should be well advanced. By 2025, renewable hydrogen would be exported. By 2030, Tasmania would have become a significant producer and exporter of renewable hydrogen while domestic use would be widely utilized. Tasmania will become self-sufficient in renewables by 2022. Beyond, around 12 gigawatts of renewable energy projects have been identified (including 8,7 gigawatts of wind power) have been identified and would enable Tasmanian renewable hydrogen to be 10% to 20% cheaper to produce compared to the other Australian states. With abundant renewable resources, access to water and adequate infrastructures (i.e.

deep-water harbour), Tasmania has not defined hard targets for the development of renewable hydrogen.

Australian Capital Territory

The Australian Capital Territory has not developed a hydrogen strategy. Nevertheless, renewable hydrogen for gas blending and mobility is considered as part of the State renewable energy plan to achieve net zero emissions by 2045 (A.C.T. Government 2020). The Australian Capital Territory suffers from drought episodes which could be an issue to produce renewable hydrogen via water electrolysis.

3.3.2   China

There is no China hydrogen strategy as such, though hydrogen research, development and applications have been supported during the five past successive Five-Year Plans. Hydrogen and fuel cell technologies are also part of the Energy Innovation Action Plan (2016–2030) among fifteen areas for technological innovation. Although China's interest in hydrogen comes from the automotive sector, the main drivers for developing its hydrogen economy are the energy security, the climate change, the air pollution, and the global competitiveness.

China is the largest producer of hydrogen globally with 21 million tons produced in 2016. Coal gasification has the largest share for hydrogen production followed by SMR, altogether accounting for over 95% of the domestic hydrogen production. The renewable energy capacity has massively increased over the past decade. With 6% of curtailment rate for solar PV and 12% for the wind power, renewable hydrogen has become a credible solution for addressing the renewable energy production waste. China has just recently started to develop its power-to-x industry. While Chinese alkaline electrolysers are well established and competitive globally, China has still a gap regarding the PEMEC technology, better suited for power-to-x application (Verheul 2019). Renewable hydrogen is expected to play a significant role in the decarbonisation of other industry sectors: ammonia and methanol production, iron and steel manufacturing and oil refining (IRENA/CNREC 2019). China hydrogen demand for the oil refinery sector could triple by 2030 (IEA 2015).

National and local policies have set ambitious and quantified targets for the deployment of hydrogen refuelling stations, fuel cell vehicles and the related hydrogen infrastructures for storage, supply, and distribution. While renewable hydrogen is considered as a storage solution to support renewable energy penetration and research programmes for power-to-x find financing, there are no quantified targets for the development of renewable hydrogen.

3.3.3   Japan

The Strategic Roadmap for Hydrogen and Fuel Cells (METI 2019) has been amended to integrate the Basic Hydrogen Strategy (2017), the Fifth Strategic Energy Plan (2018) and the Tokyo Statement (2018) to form the New Strategic Roadmap for Hydrogen and Fuel Cells has been issued in 2019. The strategy document focuses on the use and the supply of hydrogen while the production is not mentioned. For the use of hydrogen, quantifiable targets are set for the costs and the efficiency of the electrolyser technology. For the supply, a non-quantifiable target is set. The use of CO_2-free hydrogen shall be considered in a sequential manner as the processes achieve economic rationality. There is no reference or mention of renewable energy in the hydrogen strategy document.

Concurrently, the Basic Hydrogen Strategy lays out the key objectives and actions to be undertaken by 2030 to develop the hydrogen industry and eventually becoming a carbon-free society by 2050. While renewable energy technologies are mentioned as a source of CO_2-free hydrogen, brown coal combined with CCS is considered as the main path for the long-term target future picture (i.e. CO_2-free hydrogen). The strategy envisages 5 to 10 million tons per year of hydrogen consumption (METI 2017).

NEDO plays a key role in enabling the implementation of the Basic Hydrogen Strategy by providing technical and financial support to the industry actors. As of 2019, NEDO's budget for hydrogen was of USD 260 million (Ohira 2019).

3.3.4   South Korea

The third Energy Master Plan released in 2019 defines the central role of hydrogen in the future energy mix where hydrogen is considered as one of the main clean energy sources. The Hydrogen Law published in 2020 provides the legal basis to develop the South Korean hydrogen economy (IPHE n.d.). To support the application of the Hydrogen Law, the Hydrogen Economy Roadmap of Korea has been established (Government of Korea 2019). The Korean hydrogen strategy resides in the development of technologies for the use of hydrogen, mainly for the transportation and the energy sector. South Korea aims at developing its leadership in the development of downstream industries such as fuel-cells and the hydrogen infrastructure (Eastspring 2019). The development of the hydrogen industry aims at reducing the GHG and fine dust emissions. The energy self-sufficient is another driver for the development of the hydrogen industry. The National Vision for the Hydrogen Economy establishes targets to produce hydrogen vehicles and fuel cells for power generation and buildings. By 2040, 5,26 million tons of hydrogen should be produced annually. While the Vision document refers to an eco-friendly and CO_2-free supply system demand to be developed during the 2030s, there is no explicit mention of renewable energy. In the Hydrogen Supply and Prices section, water electrolysis is mentioned as a supply method, starting from 2022 and reaching 30% of the yearly total Korean hydrogen supply by 2040. The source of electricity feeding the water electrolysis is not specified. The Promotion Plan defines a transition production paradigm from "grey" hydrogen (mostly produced from petrochemical and oil refining processes) to "green" hydrogen mass imported from overseas. The Hydrogen Supply chapter provides contradictory statements. While the fossil fuel-based hydrogen extraction processes should be replaced by water electrolysis using renewable energy and hydrogen produced overseas, the same section mentions that Korea shall strive to become renewable energy-based hydrogen producing country by 2030. The Hydrogen Supply section mentions that hydrogen should be produced overseas using renewable energy and brown coal by 2030. The strategy document defines the objectives and targets for the development of the water electrolysis technology that would enable the production and supply of CO_2-free green hydrogen expending the use of renewable energy. The target is to develop a MW-class power-to-gas system by 2022 and a 100 MW-class by 2025 for commercialization purpose. The document does mention a "complex link" between the production or renewable energy and the production of hydrogen. While no quantifiable targets are provided, solar PV and power-to-hydrogen distributed systems should be developed. The strategy defines milestones for the development of the hydrogen supply from overseas. That includes the development of downstream technologies such as the infrastructure and technology to accommodate the transportation of liquefied hydrogen in the form of ammonia. Commercial and technical viability shall be demonstrated by 2025 and be fully operational by 2030.

The document identifies the Hydrogen Economy Promotion Committee as the general policy and coordination institution. It is chaired by the Prime Minister and Commissions are Ministers and private experts. An agency responsible for the implementation of the Hydrogen Economy Roadmap of Korea will be created. KRW 200 million are invested between 2021 to 2030 for launching a cross ministerial feasibility study for hydrogen. Further financial support is allocated for the development of the fuel cell vehicles, the fuel cells for power application and the downstream technologies.

South Korea aims at creating 420.000 jobs (75% in the automotive industry), KRW 43 trillion of GDP, 10,4 million TOE of hydrogen-derived energy, 27,28 million tons of GHG reduction and 55.949 GWh of power generation.

The decarbonisation of industry sectors such as the oil refining and the steel manufacturing could create a hydrogen demand of 2,8 million tons by 2030 and 4,2 million tons by 2050 (McKinsey & Company 2018).

3.3.5   Taiwan

While the same research criteria were used to identify grey literature concerning Taiwan hydrogen policy framework, limited results were yielded. Taiwan has established a Strategic Planning of Promoting Hydrogen and Fuel Cell which defines the development of 60 megawatts of hydrogen fuel cells by 2025 (R. Wang 2017). Taiwan has listed hydrogen as one of its focuses for the development of its green energy plan (Queensland Government 2019).

Summary of chapter 3
The chapter 3 has answered the secondary questions:
- What are the ambitions of the targeted countries for the development of their renewable energy and hydrogen production capacity?

What are the underpinning policies and instruments supporting the development of renewable energy and hydrogen?

All APAC countries have well developed renewable energy policies, targets, and instruments, designed to reduce GHG emissions, meet the Paris Agreement targets, and reduce the dependency on fossil energy.

All APAC countries have identified hydrogen as a key element of the future global energy landscape. Australia, Japan, and South-Korea have developed national strategies supporting the development of a hydrogen economy.

Scientific Literature Review for Establishing the Status of the Research of the Renewable Hydrogen Framework

4

Like the practical literature review, the scientific literature review is structured on the *renewable hydrogen framework* [figure 4.1]. The scientific literature review focuses on the technologies and drivers for producing hydrogen from renewable electricity, on the hydrogen applications enabling the decarbonisation of the relevant industry sectors and on the underpinning policy instruments.

To conduct the scientific literature review, the author used three main sources of information: ScienceDirect®, ResearchGate®, and Google Scholar®. The researcher has used open source and free scientific publications. The research criteria consisted on:

– The publications that have been issued during 2016 and onward. In some cases, when little results were yielded, earlier publications were considered.
– The headers of the scientific literature chapter were used as keywords.

Some additional scientific publications have been identified in the list of references of the results yielded from the scientific literature research criteria.

4.1   The Link Between Renewable Energy and Hydrogen

As the share of renewable electricity in the power mix increases, it becomes increasingly difficult to match the demand with the generation. Moreover, the construction of solar PV parks and wind farms is usually quicker than the development of the grid infrastructure. Consequently, the grid capacity may have limitations linked to transmission congestion, constraints on the local network, and balancing systems limitation. The intermittent nature of the renewable energy can lead to a back-feeding effect where more energy is produced than consumed.

M. Jaunatre, *Renewable Hydrogen*, Business Analytics, https://doi.org/10.1007/978-3-658-32642-5_4

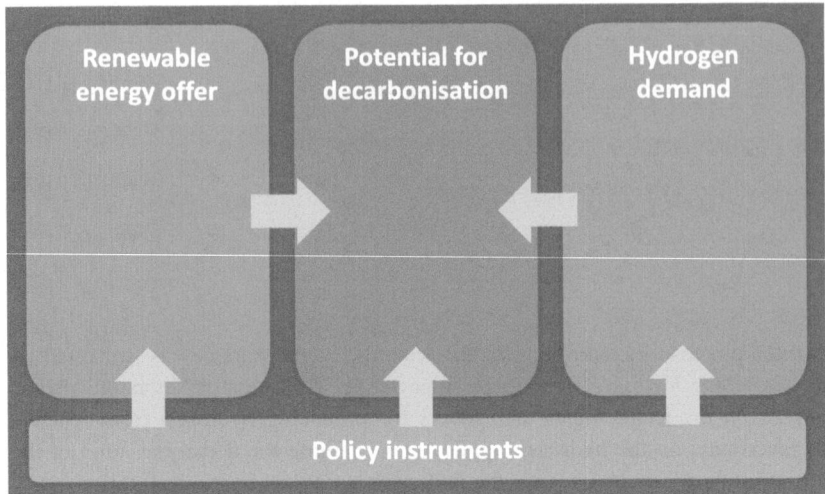

Figure 4.1 Focus areas for the scientific literature review

This typically creates voltage control issues for the grid operators (Jacobs 2016). This phenomenon in its extreme form is called the "duck curve". It was observed in California where there was an excess of solar PV generation during the day as the demand for energy is low. Towards the end of the day, solar PV generation fades while electricity demand rises sharply. This steep demand ramp-up causes issues for the grid operators as large amount of dispatchable power is then needed (Torabi, Gomes, and Morgado-Dias 2018).

In this context, different strategies are implemented to mitigate and control the effects of the mismatch between the demand and the generation.

The simplest strategy is the curtailment of the excess renewable electricity generation. Although this strategy is remarkably effective to control the effects of excess generation to the grid, it also comes with disadvantages. An operator of a renewable energy power plant may lose revenues depending on the curtailment compensation measures in force. Alternatively, compensation will be paid for a power that is not used. Moreover, the curtailment does not help at resolving the issue of the evening demand ramp-up as the curtailed energy is lost when not stored (Bird et al. 2016). Different methods for renewable energy curtailment are employed depending on the renewable energy regime in place. "Peak-shaving" curtailment is determined as a fixed percentage of the renewable power peaks.

The "renewable power-following" is done at a constant fixed percentage. The curtailment can finally be "load-following" where the curtailment rate is defined at a fixed level of the electrical load (Arabzadeh, Pilpola, and Lund 2019).

A global review of the wind and solar energy curtailment showed the growing share of wind and solar energy generation is responsible for the increasing levels of curtailment. In 2013, China had 77.16 gigawatts of wind power capacity producing 142 terawatt-hours of energy for which 16.23 terawatt-hours was curtailed (Bird et al. 2016). The highest levels of curtailment were observed in the north-eastern regions of China. In these regions, there is a high concentration of wind generation capacity associated with a low energy demand and an insufficient transmission capacity. Moreover, the lack of flexible peak power generation capacity and the absence of demand-site management hinder the integration of additional intermittent renewable power generation in the electricity mix.

In 2012, Japanese utilities still held their rights to curtail wind and solar energy up to thirty days per year (8% annual) before implementing curtailment of other power producers and suppliers (i.e. combustion and baseload power plants) (Bird et al. 2016).

Where the grid operations are managed based electricity market signals, automated methods for enabling curtailment are more efficient than manual curtailment processes. They can reduce the overall curtailment levels. It also allows solar and wind power producers to choose not to generate if prices are negative (Bird et al. 2016).

An alternative or complementary strategy to the curtailment consists on storing the electricity when it is not needed and have it available when the demand increases or when the grid frequency needs to be stabilized. The energy storage provides additional flexibility to the electricity system and contribute to mitigate curtailment. It is a complex process and different storage technologies are available with very distinctive characteristics (Luo et al. 2015). Depending on the location, the performance and the costs, some storage technologies will be better suited than others (Lewandowska-Bernat and Desideri 2017).

Renewable energy under the form of electricity can be converted to chemical energy in the form of hydrogen through the water electrolysis process (Carmo and Stolten 2018). Hydrogen can be stored in large quantity. The storage duration can range from hours to months. Fuel cells can then discharge energy into the grid at power rating within a range of seconds to hours.

The process of converting renewable electricity to hydrogen through the water electrolysis process finds other applications than the storage of energy. Called power-to-gas, this technology finds several applications that can be considered for a deep decarbonisation of industry sectors using fossil fuel-based hydrogen as feedstock.

4.2   The Review of Hydrogen Production Processes

There are several processes to produce hydrogen. The scope of the research work is limited to the renewable source powered electrolysis process and the fossil fuel-based reforming processes [Processes squared in red in figure 4.2].

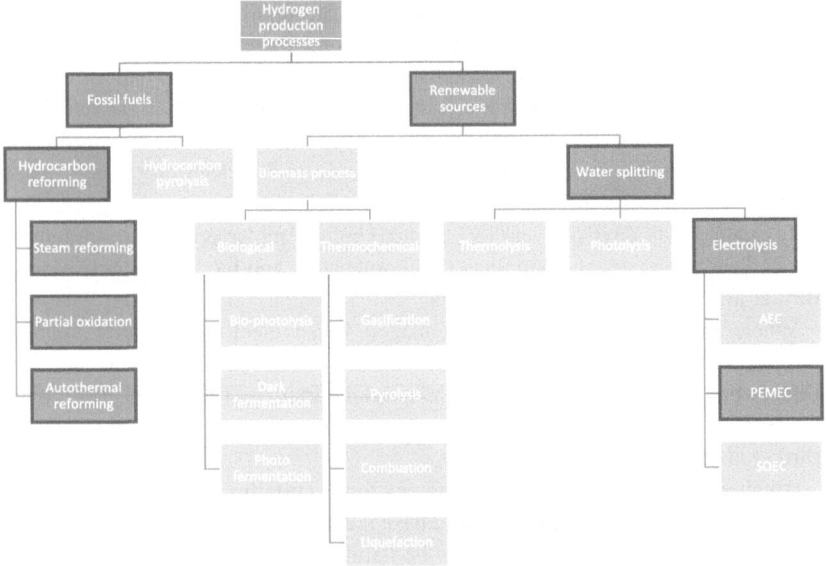

Figure 4.2 Hydrogen production processes (Shiva Kumar and Himabindu 2019)

The process of the hydrogen production and its associated costs are determined by the design of the production system, its operation characteristics and the quantity and type of losses encountered during the production and the delivery to the end user. The whole hydrogen lifecycle shall be considered from the production, storage and to the distribution (Thanapalan et al. 2012). In case of hydrogen produced from renewable energy sources, the cost of hydrogen depends from the distance between the renewable energy source to the end user.

4.2.1   The Production of Hydrogen from Fossil Fuels

Steam Methane Reformation
Technology
The SMR process is the most common hydrogen production process. The process is mature and has been used in the bulk chemical industry for decades (Van Gerwen, Eijgelaar, and Bosma 2019). The hydrogen is produced from two sequential processes: the steam reforming ($CH_4 + H_2O \rightarrow CO + 3H_2$) following by the water-gas-shift ($CO + H_2O \rightarrow CO_2 + H_2$) (Rosetti 2007). The process efficiency ranges from 65–75% (Staffell et al. 2019).

The SMR process is either integrated into an industrial process or can also be operated as a standalone system for production of merchant hydrogen. The SMR technology coupled with CCS aims at carbon neutrality (Collodi et al. 2017). However, while the CCS system would incur additional costs, its efficiency has not been demonstrated. At best, 10% of the CO_2 would still be released in the atmosphere (Fargere et al. 2018). Should the SMR and CCS be carbon neutral, it would compete with the production of hydrogen from electrolysis.
Cost
The SMR produces hydrogen at a cost of $ 1,30–1,50 kg^{-1} (Davis et al. n.d.; van Leeuwen and Mulder 2018). Adding CCS to an SMR based H_2 plant results to an increase in the LCOH between € 0,021 to € 0,051 Nm^3 H_2 (Collodi et al. 2017).
Coal Gasification
Technology
The coal gasification process is another mature process. It is mainstream in China due to the higher prices of natural gas and the abundance of the coal resource (Deng et al. 2010). It is also used in Australia. The process consists on the gasification of carbonaceous materials where both pyrolysis and combustion products are formed within the gasification chamber. The pyrolysis consists on an anaerobic conversion of the feed material (i.e. coal) into volatile compounds using heat. The volatile compounds are then combusted in the presence of limited oxygen supply. Steam is injected working as a reactant and as further source of hydrogen ($C + H_2O \rightarrow CO + H_2$, then $CO + H_2O \rightarrow CO_2 + H_2$). This process works at temperatures greater than $1.000^\circ C$ (Simons and Bauer 2011). The efficiency of the process ranges from 45%–65% (Staffell et al. 2019).
Cost
The coal gasification produces hydrogen at a cost of $ 1–1,29 kg^{-1} (IEA 2019; Ramsden et al. 2013).

Oil Refining by-Product
Technology
In oil refining facilities, hydrogen can be produced on-site from catalytic steam reforming of naphtha (C_nH_{2n+2} + n H_2O ↔ n CO + (2n + 1) H2) (Rosetti 2007), partial oxidation and catalytic cracking. Hydrogen is also produced as by-product of other manufacturing processes (i.e. hydrogenous off-gases) (Deng et al. 2010; Matijašević and Petrić 2016; Rabiei 2012).
Cost
Depending on the volume of oil processed, the on-site hydrogen production in oil refineries is cost-competitive with the SMR and the coal gasification.

4.2.2 The Production of Hydrogen from Renewable Energy with the Power-to-Gas Technology

The Power-to-Gas Technology
The core of a power-to-gas system is the electrolyser (van Leeuwen and Mulder 2018). In an electrolyser, the electricity splits the water into hydrogen and oxygen. There are three main water electrolysis technologies: the alkaline electrolysis cells (AEC), the polymer electrolyte membranes (PEMEC) and the solid oxide (SOEC). These three technologies are covered in length in the scientific literature (Kær, Søren Knudsen; Al Shakhshir 2016; Kwasi-Effah, Obanor, and Aisien 2015; Schmidt et al. 2017; Shiva Kumar and Himabindu 2019; Staffell et al. 2019).

The AEC has been in commercial use since the 1920s. This is a mature technology. The key disadvantage of this system is its inability to produce high pressure hydrogen. The hydrogen density being low, it needs to be pressurized for transportation purpose. A compressor is needed which adds costs to the system. It also increases the space requirement. Another disadvantage of the AEC is that it runs on low density currents which is not adequate with some of the renewable electricity sources.

The PEMEC is less mature than the AEC but has some interesting characteristics. It is more efficient than the AEC. The design is more compact and can be scaled up in modules. The PEMEC can operate in part-load and in overload conditions. The CAPEX and operating costs are however higher than for the AEC.

The SOEC is the least mature technology and it is not yet in commercial use. The SOEC technology has a major advantage as it can operate both ways (i.e. water electrolysis and reverse water electrolysis). It is therefore very flexible. However, it operates at higher temperatures than the AEC and PEMEC which creates opadvanced wear and tear of expensive equipment.

In a cost and performance elicitation study of the water electrolysis, the PEMEC technology is considered as the leading technology in the mid-term due to its superior characteristics for intermittent operations, consistent with the renewable electricity energy supply (Schmidt et al. 2017).

A PEMEC electrolyser is constituted of cells that kept together in a stack. Water is pumped through the stack. Direct current is circulated through the system. The electrolytic process splits the water H_2O into O_2 and H_2 at the anode. As ionised H^+ migrates through the membrane, it recombines into H_2 with electrons at the cathode. O_2 and H_2 are then collected out of the stack. Depending on the power-to-gas plant configuration and the end-user requirements, a compressor and a storage system may be required. The system will be equipped with a balance-of-plant. Transportation and distribution systems complete the power-to-gas system. The main feedstocks for a PEMEC electrolyser are water and electricity (van Leeuwen and Mulder 2018).

Additional technical information is discussed in the chapter "Empirical research for establishing the potential of renewable hydrogen for decarbonising the industrial sectors using fossil fuel-based hydrogen within the APAC markets".

The Power-to-Gas Economics

The cost elements and revenues of the power-to gas technology are represented in the figure 4.3.

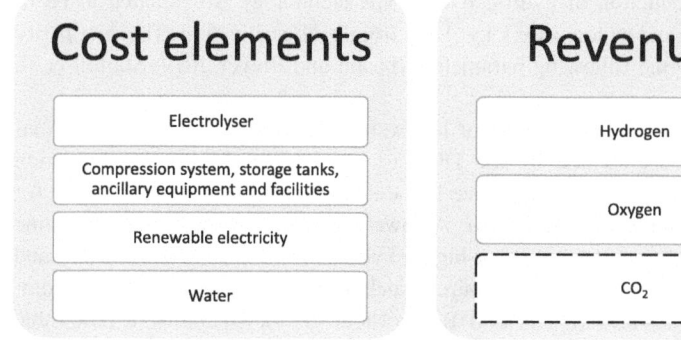

Figure 4.3   The costs elements and revenues of the power-to-gas technology

Among the cost elements, the electrolyser system is estimated to be in a range of €1.100–1.800 kW_e^{-1} for a PEMEC electrolyser. The electrolyser stack is responsible for up to 60% of the CAPEX (IEA 2019). Storage tanks and facilities are estimated to range from €20–490 Nm^{3-1}. Depending on the plant configuration and the end-user needs, the CAPEX for the compression system is estimated in the range of €144–18.500 kW_e^{-1}. Building the power-to-gas is estimated to be 30% of the CAPEX for the installation. The electricity and the water are other important cost elements depending on the market and the geographical location. The need for water is estimated to be of 10 litres kg^{-1} H_2 (van Leeuwen and Mulder 2018).

Hydrogen is the main revenue element. The oxygen market is relatively small. Currently, power-to-gas facilities release the oxygen in the atmosphere. However, in case there is demand, it could increase the competitiveness of renewable hydrogen. For instance, it can be used for the oxygenation of the sludge in water treatment plant (van Leeuwen and Mulder 2018; Staffell et al. 2019).

Currently, estimates the cost of hydrogen from wind power at about €5–6 kg^{-1} H_2 and from €3–5 kg^{-1} H_2 from solar PV (Thanapalan et al. 2012). However, production quantity and purity have an impact on the cost of hydrogen. Large-scale production and lower purity requirement can drive down the cost of renewable hydrogen production to a price range as low as €1.5–2.5 kg^{-1} H_2 (Glenk and Reichelstein 2019).

When a power-to-gas facility is integrated to a renewable energy system, the investment in the electrolyser brings an economic value when the sale value of the hydrogen is higher than the LFCH production (Glenk and Reichelstein 2019). However, the power-to-gas facilities are not profitable under current market conditions. The production of hydrogen via SMR technology is estimated to be of €1,25 kg^{-1} H_2 and as low as €1 kg^{-1} H_2 from coal gasification. The cost parity is influenced by the following parameters (Fasihi and Breyer 2019; Staffell et al. 2019).

The first parameter is the cost of renewable electricity. The Hydrogen Analysis (H_2A) model created by the DOE to estimate the hydrogen production costs showed that the electricity price is the single most impactful parameter for hydrogen production (Jacobs 2016). A lower price of electricity has the same effect as the decrease of CAPEX or higher hydrogen revenues (van Leeuwen and Mulder 2018). Renewable energy sources such as wind or solar PV are cost competitive energy sources (Glenk and Reichelstein 2019). Operating a renewable power-to-gas facility in a regulated power market makes enables the comparison of electricity prices in the day-ahead and intraday markets making possible the decrease of hydrogen production costs (Jacobs 2016; van Leeuwen and Mulder 2018).

The second parameter is the cost of fossil fuel feedstocks. The higher the natural gas to crude oil prices ratio is, the quicker the price parity is reached. In case of Japan, a 3% to 8% increase should be added to the ratio to consider the cost of regasification of the liquified natural gas (Fasihi and Breyer 2019). The expected increase of the cost of the methane feedstock will contribute to the increase of the cost of hydrogen produced via SMR. From a base price of €1,23 kg^{-1} in 2014, the price of hydrogen produced from SMR will increase to €1,54 kg^{-1} in 2024 and then to €1,84 kg^{-1} in 2034. However, a high natural gas price will contribute to a higher cost of hydrogen produced by SMR. But the higher natural gas price will also contribute to a higher system marginal costs of the electricity market. This means that a scenario with low electricity price along with a high hydrogen price is unlikely (van Leeuwen and Mulder 2018).

The third parameter is the cost of the CO_2 emissions. CO_2 emissions cost of €150 ton^{-1} would make renewable hydrogen generation cost competitive over SMR at any natural gas to crude oil price ratio (Fasihi and Breyer 2019).

This leads to the fourth parameter which is the cost of CO_2 capture. In a scenario where the cost of CO_2 emissions would increase, SMR operators would be required to capture the CO_2. This would create an extra cost estimated to be of €0,32–1 kg^{-1} H_2 (van Leeuwen and Mulder 2018).

A fifth parameter is the CAPEX of electrolysers. The learning curve of the AEC and PEMEC technologies is expected to lead to a significant CAPEX decrease over the coming years, towards a range of €500–600 kW_{el}^{-1} (Glenk and Reichelstein 2019).

The technology efficiency improvement is the sixth parameter. A higher HHV of up to 75% should be achieved by ongoing research and development of the electrolytic hydrogen production technologies (Jacobs 2016; van Leeuwen and Mulder 2018). However, its impact is limited compared to the CAPEX parameter.

A seventh parameter is the annual operating hours of the power-to-gas installation. The intensive use of the power-to-gas plant is required to cover the high electrolyser CAPEX expenditure (Glenk and Reichelstein 2019). A power-to-gas installation should run at least 2550 hours $year^{-1}$ to be competitive in a 2030 projected European market conditions (van Leeuwen and Mulder 2018).

Finally, the eight parameter is the revenues from hydrogen. The production from renewable electricity could justify the payment of a bonus compared to fossil fuel-based hydrogen (van Leeuwen and Mulder 2018).

All these parameters can be influenced by policy instruments.

4.3   The Use of Fossil Fuel-Based hydrogen in the Industry

4.3.1   The Production of Ammonia

Ammonia has been produced since the 1020s based on the Haber-Bosch process (IEA 2017). The process consists on synthesizing reacting atmospheric dinitrogen with hydrogen in the presence of iron at high pressures and temperatures ($N_2 + 3H_2 \leftrightarrow 2NH_3$). Originally designed to provide raw material for explosives to be used in weapons, the Haber-Bosch process has changed the agriculture and industry paradigm. Over 80% of the ammonia produced with the Haber-Bosch process is used for fertilizers manufacturing (Erisman et al. 2008). Ammonia is also used as a hydrogen storage medium, chemical feedstock, clean burning fuel for transportation and power generation sectors, and as refrigerant fluid. The feedstock hydrogen is most produced from SMR of natural gas. The Haber-Bosch process uses around 3%–5% of the world natural gas production, releasing 12 giga tons yr^{-1} of CO_2 for a 400 million tons yr^{-1} of ammonia produced (L. Wang et al. 2018).

4.3.2   The Production of Methanol

The industrial production of methanol can be traced back to the 1920s. Methanol is synthetized from syngas, a mixture of hydrogen and carbon monoxide together with a small amount of carbon dioxide ($2H_2 + CO \rightarrow CH_3OH$; $CO + H_2O \rightarrow CO_2 + H_2$; $3H_2 + CO_2 \rightarrow CH_3OH + H_2o$). The reaction is done in the presence of metallic catalysts (commonly $Cu/ZnO/AL_2O_3$) under harsh pressure (5–35 MPa) and temperature (200–400 ^{O}C) conditions. Methanol production is driven by the growing demands for its derivatives: olefins, formaldehyde, acetic acid, dimethyl ether (DME), aromatics, gasoline, methyl tert-butyl ether (MTBE). China is the largest methanol producer globally with over 60% of the global production (Xu et al. 2017).

70% of the methanol produced in China is coal based (De Pee et al. 2018). However, the hydrogen used in the methanol synthesis process can also be from SMR of natural gas. Syngas as waste-feed from ammonia production can also be used (Andersson and Grönkvist 2019).

4.3.3   The Oil Refining

In oil refineries, hydrogen is used in complex chemical processes such as hydro-cracking, hydro-treating, and hydro-desulfurization processes enabling the distillation of crude oil into a wide range of products. The amount of hydrogen required for processes depend on the hydrogen content of the feed and products. It also depends on the number of heteroatoms to be removed such as sulphur or nitrogen. Some oil refining processes produce purge gas rich in hydrogen (by-product). The by-product hydrogen can be treated in hydrogen recover unit (HRU) and reused in the oil refining process (Rabiei 2012).

More stringent regulations for desulphurization has led to an increase of the demand for hydrogen. The oil refineries are equipped with hydrogen production units (HPU) usually consisting on SMR plant of natural gas (Brau 2013).

If the demand from the oil refinery cannot be covered by the HRU and the HPU, merchant hydrogen is procured (EPA 2008).

4.3.4   The Manufacturing of Steel

The most utilized steel manufacturing process is the BF-BOF route. In this process, the iron ore is reduced by carbon monoxide in the blast furnace (BF). Coke and coal are used as reduction agents. The reducing agents react with the oxygen (BOF) to produce the carbon monoxide. The carbon monoxide reduces the iron ore to metal iron producing CO_2 ($Fe_2O_3 + 1{,}5\ C \rightarrow 2Fe + 1{,}5CO_2$; $Fe_2O_3 + 3CO \rightarrow 2Fe + 3CO_2$). The resulting of the process is molten iron, slug, and blast furnace gases.

Another competing process is the DRI-EAF route. The DRI process consists on the direct reduction of the iron ore by hydrogen ($Fe_2O_3 + 3H_2 \rightarrow 2Fe + 3H_2O$). The process produces water and a solid sponge iron. The solid iron sponge is then melted by the EAF process to produce steel (Otto et al. 2017). The hydrogen currently used in the DRI-EAF route is produced from SMR of natural gas, and though the DRI-EAF process is less CO_2 emitting than the BF-BOF process, it is still not carbon neutral as such (Åhman et al. 2018).

4.4 The Substitution of Fossil-Fuel Based Hydrogen by Renewable Hydrogen

In a context of increasing curtailment episodes, negative market prices and competitive and mature energy storage alternatives, the power-to-gas technology becomes a relevant candidate to secure additional revenue streams for renewable energy projects developers. The industrial sectors working on baseload operation to boost production and profitability require reliable sources of hydrogen. The demand is mostly fulfilled by fossil fuels which have severe sustainability and economic limitations. Currently, 45 to 65 Mt year^{-1} hydrogen is produced globally as feedstock for chemical and petrochemical industries. This is equivalent to about 1% of the global energy supply. 50% is produced from SMR, 30% from partial oxidation of crude oil, 18% from coal gasification and the remaining 4% from water electrolysis (Staffell et al. 2019). The supply of renewable hydrogen would enable the decarbonisation of the non-energetic use of fossil fuels in the petrochemical industry (ammonia, ethanol, oil refining) and in the future steel industry (Fasihi and Breyer 2019). However, long equipment lifetime such as SMR and investment cycles indicates that change will be slow (Staffell et al. 2019).

The power-to-gas technology provides flexibility in three dimensions (van Leeuwen and Mulder 2018). The first dimension is the time. The time flexibility enables adapting the timing of using the electricity and producing the hydrogen, for instance when the electricity prices are negative. The second dimension is the location. Instead of transporting the energy in form of electricity, the energy can be transported in a chemical form with hydrogen. Existing pipelines and shipping lines can substitute capital-intensive electricity infrastructures. The transportation of energy in chemical form reduces losses occurring during the transportation process. The renewable energy can then be provided to end-users located far away from the renewable energy production location. The third dimension is the end-use flexibility. With the change of energy vector, the renewable energy can be used in industry sectors that cannot be easily electrified (i.e. fertilizer manufacturers, steel manufacturing, oil refinery). This is called sector-coupling. The hydrogen can be used as such (i.e. in fuel cells) or converted into another substance such as ammonia or methanol (Levi and Cullen 2018).

Hydrogen produced from renewable energy sources provides a credible alternative to carbon-intensive fossil fuels. The power-to gas technology offers an excellent complementarity with intermittent renewable energy sources such as wind and solar. While the power-to-gas technology is not cost competitive with the fossil fuel-based production methods, the rapid decrease of renewable energy sources may change this paradigm. As both electricity prices and renewable power generation fluctuate over time, the investment in an integrated renewable energy and power-to-gas facility offering the ability to convert a share of the renewable electricity in hydrogen becomes relevant (Glenk and Reichelstein 2019).

4.5   The Underpinning Policy Framework

4.5.1   The Drivers and Limitations of the Current Policy Framework

Currently, the policies aiming at reducing GHG emissions are focusing on renewable energy generation. The policies aiming at increasing the affordability of the energy generally exclude hydrogen technologies due to their current high costs. Policy support for hydrogen technologies are motivated by different national priorities, whether those are the climate change, the reduction of GHG, the energy security, affordability, and the economic growth (Staffell et al. 2019). For instance, China's hydrogen policy is driven by the need of reducing the pollution in its metropoles. It also aims at supporting the economic growth through the development of hydrogen-based downstream applications (i.e. fuel cells). For Japan, the main hydrogen policy driver is the energy security. It also aims at supporting its national industries by focusing on the development of FCVs and fuel cells. The EU is considering establishing renewable hydrogen standards to support the green energy transition. Other drivers for the promotion of hydrogen technologies in the policies are the stability brought to the electricity system, the reduction of the environmental impacts by offsetting fossil fuel-based technologies, the development of new carbon-free technologies and markets with their positive fallouts on the economic growth and the employment.

A global and coordinated development of an economic and a regulatory framework to support the development of the hydrogen produced by water electrolysis is missing (Lewandowska-Bernat and Desideri 2017; Thanapalan et al. 2012). Focused, predictable and consistent energy policy are the basis of successful innovation (Staffell et al. 2019). Unpredictable and suddenly changing policies undermined the confidence of the investors. The phenomenon has been observed in Spain for instance where the radical stop of fit in tariffs for solar PV triggered a halt in the development of renewable energy in the middle of the 2000s. This is also relevant for the development of the hydrogen technologies. The policy mechanisms supporting the development of hydrogen technologies should be defined and communicated well in advance. That way, the hydrogen technologies could mature, and the costs could decrease significantly, like what happened for the wind and solar PV technologies.

In liberalized electricity markets, four main policy instruments have supported the development of renewable energy projects:

- A price for the electricity production regardless of the market price such as the feed-in tariffs.
- A fixed amount on top of the electricity market price such as a feed in premium or as a production tax credit.
- A grant for the initial investment, a tax credit, or an accelerated depreciation scheme.
- Renewable portfolio standards acting as a mandate to increase the renewable electricity production.

These policy instruments have been highly effective at decreasing the cost of renewable energy to a point that the LCOE of a large-scale renewable energy plant is lower than gas or nuclear (Joos and Staffell 2018). Policies supporting the development of renewable energy generation have an indirect positive effect on the renewable hydrogen business case as the renewable electricity prices decrease.

However, due to an almost zero marginal costs of renewables, the intermittent nature of renewables and the interplay between the price volatility and renewable technologies, the policies that have been successful to support the development of renewable energy will become less successful as the share of renewable in the energy mix increases. As the prices of the electricity market decreases, the investors will not invest in a renewable energy project without enough guarantees on returns (Blazquez et al. 2018).

4.5.2   The Recommendations for Policy Development

The policy instruments that have been used to develop the global renewable energy generation capacity can also be utilized to develop the renewable hydrogen production through water electrolysis. Moreover, specific policy instruments for renewable hydrogen can be considered.

The creation of a so-called "green" hydrogen market would result in a premium for the selling price of hydrogen produced from renewable sources (van Leeuwen and Mulder 2018). This would imply the creation of certification system for the hydrogen produced from renewable energy.

In a policy framework where the renewable energy producer gets a premium when the electricity is fed to the grid (i.e. feed-in tariffs), there is little incentive to utilize the renewable energy for producing hydrogen. In this context, the reduction or withdrawal of the support provided for renewable electricity generation fed into the grid would support the economic viability of hydrogen produced from renewable energy sources (Glenk and Reichelstein 2019).

Straight rebate or investment tax credit for investment in electrolytic hydrogen production technologies would also support the emergence of renewable hydrogen production (Glenk and Reichelstein 2019).

If renewable hydrogen is used for storage of renewable energy, power-to-gas facilities could also be granted rebates in connection with battery storage (Glenk and Reichelstein 2019).

Policymakers could also consider the positive impact of higher fossil fuel prices on the cost competitiveness of hydrogen produced from renewable energy sources (Glenk and Reichelstein 2019).

Finally, investment programs for hydrogen related technology development is another relevant example of policy instrument (Thanapalan et al. 2012).

Summary of Chapter 4

The chapter 4 has answered the secondary question:

- What is the hydrogen demand of industry sectors using fossil fuel-based hydrogen as feedstock?

Without answering the question from a quantitative point of view, the chapter 4 has shown that hydrogen is a key feedstock for industrial processes such as the ammonia and methanol production, the oil refining, and the steel manufacturing. Hydrogen feedstock is produced from CO_2 emitting processes such as the steam reforming of natural gas, the coal gasification, and the oil refining by-products.

Currently, the renewable hydrogen production via water electrolysis is four-fold more expensive than the fossil fuel-based hydrogen production processes. It is not cost competitive unless it would be supported by specific policy instruments.

The quantitative answer is provided in the chapter 5.

Empirical Research for Establishing the Potential of Renewable Hydrogen for Decarbonising the Industrial Sectors Using Fossil Fuel-Based Hydrogen Within the APAC Markets

5

5.1 Evaluation of the Literature Review Findings Regarding the Role of Hydrogen in the Energy Transition

This chapter provides the findings from the practical and scientific literature review. The findings are addressing the renewable energy offer, the hydrogen demand, and the policy instruments elements of the renewable hydrogen framework [Figure 5.1].

5.1.1 The Renewable Energy Offer

The Renewable Energy Landscape in the APAC Markets
The documentation review shows that the countries under the research scope have established renewable energy policies and targets. The main drivers for the development of their renewable energy capacity are the decarbonisation of the energy sector and the reduction of the GHG emissions. However, the mitigation of energy dependency on fossil fuels is another driver particularly relevant for Japan and South-Korea.

The level of ambition varies among the APAC markets.

At a federal level, Australia targets are defined until 2021 and remains at the same level until 2030. An update of the targets is expected but without a defined deadline. Most of the Australian States have established their own targets. In some

---

Electronic supplementary material The online version of this chapter (https://doi.org/10.1007/978-3-658-32642-5_5) contains supplementary material, which is available to authorized users.

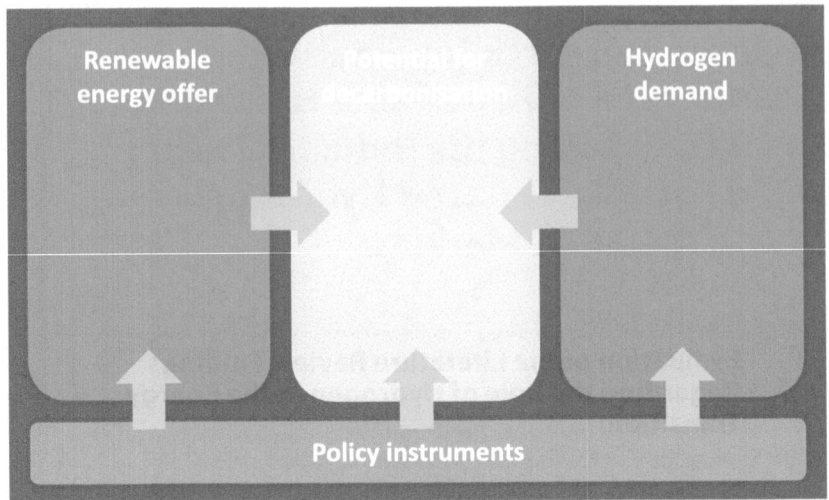

Figure 5.1 The focus areas of the documentation review

instances, the State targets are more ambitious than the Federal ones. The whole country benefits from excellent wind and solar resources. South Australia, Tasmania, Queensland, and the Australian Capital Territory have the highest renewable energy targets. Australia is an established coal and natural gas exporter. Western Australia, Queensland, South Australia, and Tasmania consider the development of renewable energy for exportation purpose.

China has the challenge to produce enough energy to sustain its economic growth while mitigating the environmental, social and health impact of its traditional coal-based industry. To address this challenge, China renewable energy targets are by far the largest in absolute numbers and the most ambitious in growth. By 2025, China plans to have over 1.000 gigawatts of combined solar PV and wind power capacity.

In 2017, Japan renewable energy capacity was fairly limited with around 8,5% of wind and solar PV generation in the final energy mix. The reduction of the dependency on imported fossil energy and the commitment to the Paris Agreement to limit the global warming are the main drivers behind Japan renewable energy development.

South-Korea is like Japan from an energy mix point of view. The limited renewable energy capacity of about 7% of the energy mix in 2019 should reach 20%

by 2030. In absolute values, this corresponds to around 40 gigawatts of combined solar PV and wind power to be built.

Taiwan has the objective to reduce the coal and oil-fired energy production and to become a nuclear free country. To reach these objectives, Taiwan has defined ambitious renewable energy growth targets, though relatively smaller than Japan or Australia in term of absolute values.

The production of hydrogen from renewable electricity
The documentation review provides information regarding the link between renewable energy and the production of renewable hydrogen.

The curtailment and the energy storage help accommodating a growing share of intermittent renewable energy production in an electricity system. The curtailment only addresses the issue of the excess generation. Depending on the legal framework, curtailed electricity is either a loss for the renewable energy project operator or it is compensated by the taxpayers. Therefore, storage offers a complementary strategy to the curtailment. The excess energy is stored and reinjected to the grid when needed. There are several technologies in place or under development, with quite different characteristics suitable for specific situations or purposes (i.e. frequency control, seasonal storage...). One storage technology consists on converting the electrical energy into chemical energy by producing hydrogen via water electrolysis. The hydrogen can be converted back to electricity via a fuel cell or a reverse electrolysis process.

Beyond the energy storage application, hydrogen produced from renewable energy can be used in other applications. This is the power-to-gas technology. A power-to-gas system provides flexibility to its operator. It can be operated on-demand, for instance when there is excess of renewable electricity available. When the renewable electricity is converted into chemical energy in the form of hydrogen, it can be transported anywhere it is needed. Finally, by changing the energy vector, renewable energy can be used in sector that cannot be electrified easily and help the decarbonisation. This is called sector coupling.

Three main water electrolysis technologies exist with different level of maturity. The PEMEC technology offers technical characteristics that are suitable with the intermittent nature of the renewable energy production. The PEMEC technology is not yet mature and therefore a decrease of the LFCH is expected along with the learning curve of the product.

The electricity and the water are other important cost elements. The higher the price of the electricity, the costlier the production of renewable hydrogen. Water can be a scarce resource in some regions of the APAC markets and could compete with other needs.

Some indirect factors impact the cost competitiveness of the renewable hydrogen production. A higher cost of the CO_2 emissions would reduce the cost competitiveness of fossil fuel-based hydrogen production compared to renewable hydrogen. Moreover, the capture of CO_2 emissions (CCS), if required, would add a cost element to the fossil fuel-based hydrogen production. Similarly, a higher cost of the fossil fuels used to produce hydrogen, for instance the natural gas, would increase the competitiveness of the renewable hydrogen production.

Long running hours of the power-to-gas system, higher merchant value of renewable hydrogen and a market for merchant oxygen would also contribute positively to the cost competitiveness of a renewable power-to gas system.

5.1.2   The Hydrogen Demand

The documentation review has shown that hydrogen has been used as feedstock in the industry sectors for decades. 45 to 65 million tons year^{-1} of hydrogen are consumed globally by petrochemical industries.

Over 95% of the hydrogen is produced from fossil fuels via SMR, coal gasification or as by-product of oil refining processes. The SMR is the most utilized process globally while the coal gasification process is mainstream in China due to the traditionally coal-based economy. Coal gasification is also widely used in Australia due to the important coal resource.

The production of fossil fuel-based hydrogen is most cost competitive compared to the production of renewable hydrogen. The production of hydrogen from fossil fuels has a cost ranging between € 1–1,5 kg^{-1} H$_2$ while renewable hydrogen production has a cost ranging between € 3–6 kg^{-1} H$_2$ depending on the renewable energy source.

Hydrogen is used in the production of ammonia. Ammonia is a key element for the fabrication of fertilizers. Ammonia can also be used in other industry sectors such as the transportation and energy sectors.

Hydrogen is used in the synthesis of methanol. China is the largest producer of methanol accounting for more than 60% of the global production capacity. 70% of the Chinese production is based on coal.

The hydrogen demand in oil refining is continuously increasing with more stringent regulations for desulphurization. Most of the hydrogen in an oil refinery is produced in-situ, either as a by-product from the refining processes or from purpose-built natural gas SMR plant. When the hydrogen demand cannot be covered by the production capacity of the oil refinery, merchant hydrogen is procured.

The BF-BOF steel manufacturing process does not use hydrogen as feedstock. However, a partial decarbonization of this steel manufacturing process can be achieved by adding hydrogen in the BF process. A full decarbonization can be achieved by replacing the BF-BOF production process by the DRI-EAF process. In the DRI process, the reduction of the iron ore is entirely done by the hydrogen. The EAF process is a process powered by electricity.

The decarbonization of the industry sectors falling in the scope of the research work is therefore possible from a technical point of view. However, from a business rationale point of view, the production of renewable hydrogen is not yet cost competitive with the fossil fuel-based hydrogen production.

5.1.3 The Policy Instruments

The renewable energy policy instruments
The documentation review shows that auctions have become the most utilized policy instruments to allocate new renewable energy projects. To win the right to build a project, the renewable energy project developers are competing on the price but also on some other criteria depending on the design of the auction system.

The auction is just one element of a wider policy system. Once the project is awarded, some other policy instruments may come into play. The feed-in tariffs are still widely utilized. However, in Europe, some projects allocated through auction do not receive any support or subsidy afterwards. This exposes the renewable energy project developers to the merchant risk and to the volatility of the electricity market. To mitigate this risk, renewable project developers secure the quality of their revenue streams by developing PPAs. Another mitigation measure is the integration of the power-to-gas technology in the renewable energy system, with the assumption that renewable hydrogen production can secure an additional and complementary stream of revenue.

With this background, the review of the renewable energy policy framework in the APAC markets under the scope of the research work have shown that China has stopped using direct financial subsidising of renewable energy production. Feed-in tariffs have been used to support the initial growth of the renewable energy capacity installation until 2019. A new mechanism is being designed and will be put into use to support the 14th Five-Year Plan.

In Australia, large-scale generation certificates (LGCs) are the main supporting mechanism for renewable energy projects. The amount of LGCs to be produced are defined on a yearly basis. Once a renewable energy project is accredited under the scheme, it can produce and sell LGCs.

In Japan and Taiwan, the feed-in tariffs are still the main policy instrument supporting the development of the renewable energy capacity. Taiwan implements a price ceiling to cap the subsidies provided.

In South-Korea, the development of renewable energy capacity is supported by a combined RPS and feed-in tariffs system.

The renewable hydrogen policy instruments
The review of the APAC markets renewable hydrogen policy frameworks has shown contrasted results. Australia, Japan, and South-Korea have clearly recognized hydrogen as a singular element of the energy landscape for the decades to come.

Australia has developed comprehensive roadmaps and strategies, at Federal and State levels. Australia's main objective is to become a global leader in the exportation of hydrogen. The rationale is to build upon Australia's existing commodities export routes to ship hydrogen abroad. However, the Australian National Hydrogen Strategy does not take a stand on the renewable origin of the hydrogen. All hydrogen production routes are considered, including the heavily GHG emitting coal gasification one. Nevertheless, Australia is the only country of the APAC markets that identifies its renewable energy capacity as a source of hydrogen production. Australia's targeted export countries are Japan, South-Korea, and China. Australia National Hydrogen Strategy is echoed in most its States by specific hydrogen roadmaps and strategy documents. Queensland, Western Australia, South-Australia, and Tasmania are the most ambitious states regarding hydrogen production and export. South-Australia and Tasmania clearly envision producing their hydrogen from renewable energy. This is different from Queensland and Western Australia. In these two States, production from fossil fuels (coal and natural gas), possibly with CCS, is the privileged scenario in the short and mid-term. Western Australia has an immense renewable energy resource that could be used for hydrogen production.

Japan and South-Korea have both established a hydrogen strategy supported by a legal framework. Both countries have set hydrogen as a top priority item of their future energy strategy. However, Japan and South-Korea are net importers of energy. In this context, the focus of the hydrogen strategy is primarily the development of downstream hydrogen-based applications, such as FCEVs, HRSs, fuel cells and hydrogen fired power plants. In a world where hydrogen would become a key energy vector, Japan, and South-Korea ambition to be world leaders in the exportation of the hydrogen technologies. For both countries, the use of hydrogen from renewable energy is considered but not before the 2030s horizon.

Both strategies put focus on developing supply chains for importing hydrogen, including the renewable hydrogen.

China is the largest hydrogen producer globally. Hydrogen has been part of the legal framework of China since the 1990s (Five-Year Plans). Also, hydrogen is one of the technological areas defined in the China Energy Innovation Action Plan. Like Japan and South-Korea, China is investing in R&D for the development of downstream hydrogen applications. National and local policies support the development and deployment of FCEVs and HRSs with ambitious targets. The power-to-gas technology is a technology identified for addressing the renewable energy penetration challenges. However, there are no target defined for this specific application.

Taiwan has identified hydrogen as a key element of its green energy plan. However, the documentation review did not allow retrieving any tangible information regarding its legal framework.

The limitations and challenges of the current policy instruments
The decarbonisation of the industry sectors in the research scope is feasible by replacing the fossil fuel-based hydrogen by renewable hydrogen. However, renewable hydrogen is not currently cost competitive. Only strong policy instruments could support reaching the cost parity with fossil fuel-based hydrogen.

There are no policy instruments that incentivize the renewable energy for the purpose of renewable hydrogen production. The review of the policy frameworks has shown that generous financial support to produce renewable electricity does not provide enough incentive to integrate a power-to-gas system. Japan, South-Korea, and Taiwan are providing substantial financial support for the renewable energy production. China, where the feed-in tariff policy has stopped with the end of the last Five-Year Plan, and Australia, using a year-ahead renewable energy certificates system, may be conducive to the production of renewable hydrogen as a way to improve the revenue stream of renewable energy project developers.

There are no direct policy instruments supporting the production of renewable hydrogen. Japan and South-Korea have defined the legal basis for the development of their hydrogen economy, mostly based on mobility, power, heat, downstream technology applications and imported hydrogen. Hydrogen is mentioned in the energy legal framework of China. Taiwan has recognised the importance of hydrogen in the future energy landscape. But none of those legal frameworks provide the legal ground to produce hydrogen from renewable energy.

Australia has defined the ambitious objective to become a global leader in the exportation of hydrogen. The Australian National Hydrogen Strategy is not directive as regard to the renewable nature of the hydrogen to be exported. However, Tasmania and South-Australia have clearly defined in their own strategies the ambition of developing renewable energy solar PV and wind farms for the sole purpose of producing hydrogen. To some extent, Western Australia and Queensland are considering producing hydrogen from renewable sources. However, those two States envisage primarily producing hydrogen from their abundant fossil fuel sources.

Other direct and indirect policy instruments can be deducted from the analysis of the cost and revenue elements of the renewable hydrogen production.

- Withdrawal of the direct financial support to produce renewable energy
- Grant for supporting the initial CAPEX investment
- Increase of the renewable hydrogen merchant value through a certificate of origin mechanism
- Increase of the carbon (emissions) price
- Increase of the fossil fuel prices
- Grant and subvention for R&D to support the learning curve of the technology

5.1.4   The Planning of the Next Steps of the Empirical Research Based on the Conclusions of the Documentation Review

The next steps of the research are the expert interviews and the survey of the study cases, providing further depth to the findings of the documentation review.

With reference to the *renewable hydrogen framework*, the interviews shall investigate the elements highlighted during the literature review and presented in the figure 5.2.

When assessing the potential for decarbonisation of the industry sectors covered by the present paper, different scenarios for the use of renewable electricity are envisaged. One scenario considers that the entirety of the renewable electricity is used to produce hydrogen. Another scenario consists on using only the curtailed electricity that would be otherwise lost. A third and last scenario consists on producing renewable hydrogen for 2.550 hours a year, corresponding to the minimum operating hours of an electrolyser to make the power-to-gas system economically viable (van Leeuwen and Mulder 2018).

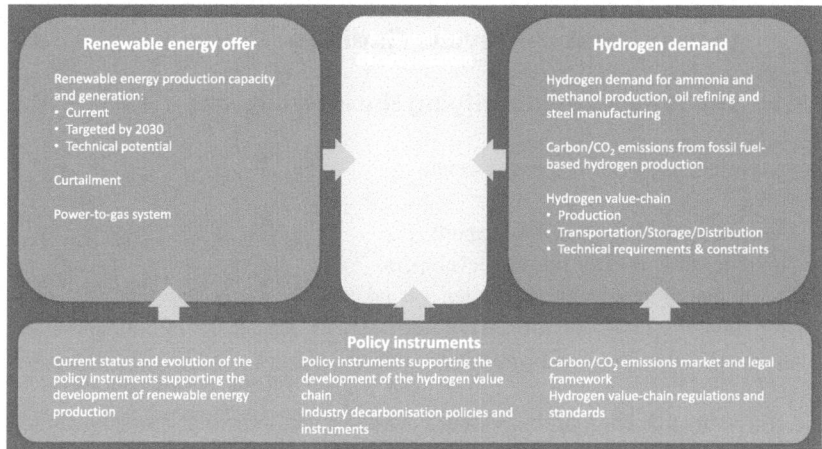

Figure 5.2 Renewable Hydrogen Framework—Focus areas identified out of the literature review

Without considering the economic aspect of decarbonising the industry sectors under the research scope, the survey consists on evaluating the impact of the renewable energy policies on such decarbonisation. In the context of this paper, the decarbonisation means replacing the fossil fuel-based hydrogen by hydrogen produced from renewable electricity through water electrolysis. The determination of the hydrogen demand serves as a reference point. The potential for decarbonisation is assessed based on the amount of renewable hydrogen that can be produced. The production of renewable hydrogen is determined based on the renewable energy production capacity resulting from the implementation of the current renewable energy policies, the planned capacity targets at the horizon 2030 and the renewable energy technical potential of each of the study case.

The outcomes of the study cases survey are expected to provide indications regarding the current and future propensity of the renewable energy policies to support the decarbonisation of industries using fossil fuel-based hydrogen.

The interviews findings and the outcomes of the study cases survey can help formulating policy recommendations as well as suggesting topics for further research.

5.2   Expert Interviews for Providing Additional Inputs to the Literature Review Findings

5.2.1   The Process of Identifying the Interviewees

Three types of interviewees are selected:

- The renewable energy policy experts,
- The renewable energy project developers,
- The potential renewable hydrogen off-takers. The latter are representatives of industry sectors mentioned earlier such as producers of ammonia and methanol, oil refiners and steel manufacturers.

The criteria for selection of interviewees are defined empirically. The subject matter experts shall justify several years of working experience in their respective professional fields. The professional fields obviously need to be relevant with the research topic.

The renewable energy policy experts are identified by scrutinizing their roles and experience in the field of:

- Intergovernmental or governmental energy and environmental organizations and agencies,
- Energy and renewable energy governmental or state agencies or bodies,
- Academic institutions focusing on energy, renewable energy, and hydrogen topics.

The subject matter experts in the field of renewable energy projects development are identified in organizations developing and building renewable energy projects, but also potentially working with renewable hydrogen projects and initiatives.

The subject matter experts for the hydrogen off-take are identified in large corporations producing ammonia, methanol, oil refining and steel manufacturing in the APAC countries chosen for the research work.

Three main sources are used to search for interviewees.

The first source is LinkedIn®. The LinkedIn® search function enables to configure searches of professionals based on their geographical location and their field of activities and expertise. The researches can be refined with keywords. When accessing a professional profile, the information provided allows screening the potential interviewee with the criteria defined above (i.e. field of expertise, years of experience, assignments, or publications of interest in scope of the research work…). When the criteria are met, the potential interviewee can be contacted via LinkedIn® InMail system.

There are a few limitations while using LinkedIn for identifying interviewees. The only resource spent by the author in the project is time. No monetary budget is provisioned to deliver the research work. This can be limiting as LinkedIn® enables only a certain number of features with a free access user profile. The LinkedIn® algorithm detects certain patterns when using the search function. As an example, when searches are done outside the geographical location or outside the LinkedIn® network of the user, the accuracy of the search results drops significantly after about ten searches. The search function is then disabled and a payment for a professional user access profile is requested. The search capabilities are redeemed weekly. After the first successful requests for connection with potential interviewees, more searches in the newly extended LinkedIn network are possible. Secondly, as a free user of the LinkedIn® professional network, the user is not allowed to directly contact the potential interviewee. Only when the connection through the platform is enabled, the user can write emails to the recipient. This can be a significant issue as potential interviewee may not be willing to accept unsolicited communication. When the user is sending a request for connection, the user is allowed for only 140 characters to motivate the connection. Lacking the ability to present oneself and not being able to provide an explanation and a purpose for the solicitation may hinder considerably the odds of the author to find interviewees. To overcome this issue, the author prepared a video explaining the topic of research, the purpose of the interview and some practical information regarding the interview. Three videos (Jaunatre 2020c, 2020b, 2020a) are made for the three targeted types of interviewees. The videos are uploaded to an online video platform. The link to the video is then added to the 140 characters introduction message allowed for a LinkedIn free user. The videos have several purposes. One is to overcome to inbuilt limitation of characters when request a potential interviewee for a connection. The other one is to create empathy and provide a human element to the request. This way, the author expects creating an element of surprise and interest, increasing the chances of success in the quest for interviewees. A semi-structured interview requires a certain degree of interest and trust between the interviewer and the interviewee (Barriball 1994). About half of the interviewees have been identified via LinkedIn®, with representatives from the policy experts and hydrogen off-takers groups. The final limitation that can be highlighted is possible discrepancy between the information published on the LinkedIn® profile and the actual curriculum vitae of the interviewee. This is mitigated by a short introduction requested at the beginning of the interviews to verify the credentials of the interviewees.

The second source of interviewees are authors of grey and scientific literature on the research topic. Some authors have been identified while performing the literature review. Authors names and contact details are usually made available in the publications. The author of this paper contacted relevant authors, mainly policy experts, via email when the address email was available on the publication or via LinkedIn® following the protocol described above. The solicitation by email is done using the same narrative as the one developed in the introduction videos (appendix 2). Two interviewees have been identified with this strategy.

The third and last source of interviewees is the professional network of the author. The author is currently employed by a company specialised in the development, construction, and operation of large-scale renewable energy projects. Moreover, the company employing the author has also launched a renewable hydrogen initiative. An energy policy expert and a renewable energy project developer are identified this way. However, by sourcing interviewees within the same organisation, the author has potentially created some limitations and biases. The strategy of the company is not representative from the all the renewable energy project developers due to its size, its global footprint, and its own business drivers. Therefore, the qualitative data collected in this setting may lack representativeness. Another renewable energy project developer may have another view on the renewable hydrogen. To mitigate this potential bias, the author has used the same interview guide and the same interview set-up for all interviewees.

As a result, fifty-seven (57) potential interviewees have been identified and contacted with the following outcome (Figure 5.3):

It is difficult to precisely define a minimum number of interviewees for a successful qualitative data gathering research. The literature defines the moment where enough data is gathered as the point of data saturation (Hagaman and Wutich 2017). Different figures are proposed in the literature, ranging from five to twenty-five (Townsend 2013) up to twenty to forty (Hagaman and Wutich 2017). Other source (Guest, Bunce, and Johnson 2006) suggests that at least twelve interviews in a homogeneous group are needed to reach the point of data saturation. The author was originally considering interviewing fifteen persons. However, given the timeframe of the research work and the virtual setting, the target has not been reached. In total, ten interviews have been confirmed. This implies that the point of data saturation may not have been reached. Furthermore, despite of promising contacts and discussions, none of the identified Chinese interviewees were able to commit for an interview.

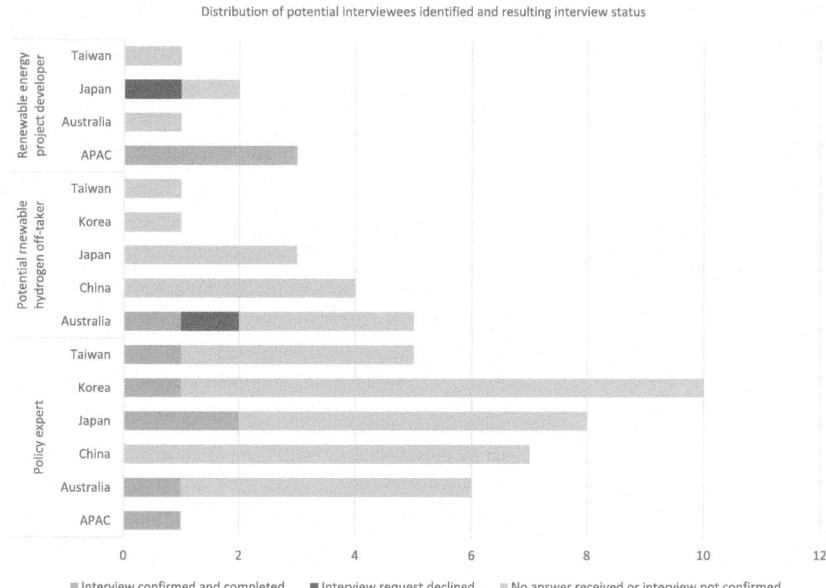

Figure 5.3   Distribution of interviewees and status of interviews

5.2.2   The Profiles of the Interviewees

According to the interviewee protocol, the interviewees remain anonymous. However, their names and respective curriculum and credentials are submitted for review to the thesis supervisor for validation of the qualitative data gathering process.

The table 5.1 provides an overview of the confirmed interviewees profiles, their area of expertise and in which geographical area this expertise applies.

The interviewee #10 has a significant experience in the development of large-scale renewable energy projects in the Northern Europe. The interviewee #10 has also recently actively participated in promoting the development of the Dutch renewable hydrogen sector together with other industry actors such as fertilizers manufacturers using hydrogen as feedstock. The Dutch renewable hydrogen initiative is the most advanced and developed globally. The author uses the outcomes of the interview #10 to set a benchmark when reviewing the findings of the other interviews.

Table 5.1 Confirmed interviewees profiles

#	Interviewee type	Area(s) of expertise	Geographical area
1	Potential renewable hydrogen off-taker	Integrated renewable energy and renewable hydrogen projects for industrial and mobility applications	APAC, Australia
2	Renewable energy policy expert	Solar PV projects development Renewable energy policies and business development	South-Korea
3	Renewable energy policy expert	Renewable energy policies Governmental renewable energy policy instruments design	Taiwan
4	Renewable energy policy expert	Renewable energy policies and development mechanisms	Japan
5	Renewable energy policy expert	International energy policies Natural gas business development	Japan
6	Renewable energy policy expert	Renewable energy policy development Hydrogen business development	Australia
7	Renewable energy policy expert	Global petrochemical and hydrogen value chains	APAC
8	Potential renewable hydrogen off-taker	Fertilizers R&D and production Ammonia and hydrogen value chains	Australia

(continued)

Table 5.1 (continued)

#	Interviewee type	Area(s) of expertise	Geographical area
9	Renewable energy project developer	Regulatory framework, support mechanisms and renewable energy policies	APAC, Japan and South-Korea
10	Renewable energy project developer	Large-scale renewable energy project development (offshore wind) Integrated renewable energy and hydrogen systems business development	Northern Europe (used as benchmark)

5.2.3   The Design and Contents of the Interview Guides

Three different interview guides are established to fit the three different types of interviewees. The interview guides are displayed as appendixes 3, 4, and 5 of the present document.

The first part of the interview guides is common for the three different types of interviewees. In a first section, it comprises a short introduction of the author and a description of the research topic. In a second section, the author explains the purpose of the research providing additional background and context for the research topic. It identifies the four main elements of the *renewable hydrogen framework* as displayed in the figure 1: the development of renewable energy projects, the demand for hydrogen which is currently produced mostly from fossil-fuels, the potential for decarbonisation using hydrogen from renewable source and the underpinning regulatory framework covering those elements. As the background for the research is presented to the interviewee, a third section describes the rationale of the semi-structured interview and the intended use of the qualitative data collected during the interview. The fourth section provides practical details regarding the interview itself, such as the invitation for the interview, the duration, the various settings proposed for the interviewees (i.e. virtual meeting, phone call and email) and the contact details of the interviewer. Finally, the common part of the interview guides ends with a fifth section defining the confidentiality procedure the interviewer intends to implement.

The first common section for all interviewee types aims at providing enough context, background, trust, and authenticity to the potential interviewees, setting solid ground rules for an efficient and in-depth discussion (Adams 2015).

The structure of the second part of the interview guides is also designed according to the *renewable hydrogen framework*. The interview guide is designed to facilitate an hour interview and its first section sets for an introduction. As explained in section 5.2.1, the author needs to confirm that the interviewee corresponds to the profile reviewed during the interviewee identification phase. The next sections are tailored to reflect the areas of expertise and interest of the three (3) types of interviewees as shown in the table 5.2.

Table 5.2 Structure of the interview guides

Interview guide for renewable energy policy experts	Interview guide for renewable energy project developers	Interview guide for potential renewable hydrogen off-takers
Introduction (common)		
Renewable energy policy instruments (common) *Four questions*		Sustainability and environmental management framework (tailored) *Two questions*
		Hydrogen as feedstock (tailored) *Nine questions*
Renewable hydrogen (tailored) *Eleven questions*	Renewable hydrogen (tailored) *Thirteen questions*	Renewable hydrogen (tailored) *Eight questions*
Concluding remarks (common)		

5.2.4   The Organisation and Realisation of the Interviews

The first requests for interviews have been sent end of November 2019 until mid-December 2019 with an objective of performing the interview by end of January 2020 in accordance with the original timeline of the research project.

Once confirmed, the interviews have been completed between the 21st of January 2020 and the 14th of February 2020. Any developments or information relevant to the research topic that may have occurred beyond that date are not covered by the research.

The interviews have consisted of an hour discussion following the questions documented in the interview guides. Eight interviews have been conducted via Skype®, allowing the recording of the discussions. Two interviews have been conducted by phone. The interviews have been transcribed into writing format. The documented transcriptions were submitted to the interviewees for comments and for reference. Three of the interviewees gave minor comments as well as expressing appreciation for the interview process.

The interview transcripts were used for qualitative data analysis.

5.2.5   The Main Findings of the Interviews

The interviewee #10 is working in the Dutch market. The Netherlands are at the forefront of the renewable hydrogen development for industry applications. The strategy of the interviews is to use the data gathered in the Dutch market as a benchmark for the data gathered in the APAC markets.

The Dutch market benefits from a combination of factors that has led to the positive development of the renewable hydrogen topic.

- The policy instrument to allocate offshore renewable energy projects is a zero-subsidy auction. This exposes renewable energy project developers to the merchant risk. Renewable energy project developers are pushed to develop additional revenue streams to mitigate this risk.
- Though grid capacity is not yet an issue in the Netherlands, reinforcement of the grid will be needed soon to accommodate large intermittent input of offshore wind electricity. Renewable hydrogen as an energy storage medium could play a role in providing flexibility and reduce the need for grid infrastructure improvement.
- The Dutch government has ambitious decarbonisation goals for the industry. For decarbonizing the industry, large quantity of renewable hydrogen will be required. This pushes industry actors to develop ambitious sustainability strategies and rethink their industrial processes.
- The Dutch government has defined ambitious goals for the development of its renewable energy production capacity and so far, it has been followed by actions.
- The Dutch government sees hydrogen as a key contributor to the energy transition.

- In addition to the EU ETS (Emissions Trading Systems), a carbon floor price is under discussion between the Dutch authorities and the industry.
- There is already a big hydrogen market where 800.000 tons of hydrogen are produced yearly. Scale is key for reduction of costs.
- This hydrogen demand is concentrated on petrochemical clusters which are closed to the landing points of offshore wind farms export cables.

The renewable hydrogen Dutch model has some limitations.

- The definition of the term "renewable hydrogen" is not officially established. This has an impact on how the SDE + + (i.e. Dutch technology neutral subsidy system) subsidies will be allocated.
- Forecast data models used as reference to design policies are based on an electricity mix still dominated by fossil energy. This makes the business case for renewable hydrogen not viable.
- The current policy framework for battery storage needs to be updated to unlock the potential of storage technologies. Today a fee must be paid when electrons are stored and then when the electrons are put back to the grid.
- The current (zero-subsidy bid) policy framework is not sustainable in the long term for renewable energy projects developers and investors as they are fully exposed to merchant risk. Uncertainty may reduce investments which may in turn reduce development of renewable energy projects.
- The renewable hydrogen is not yet supported by a clear policy instrument. This is however under discussion and should materialize in short term.
- The policy instruments for development of large renewable energy projects and those developed for renewable hydrogen are still "decoupled". However, the Dutch authorities are seriously considering integrated tender systems. Intense lobbying is currently undertaken by the renewable energy projects developers to influence policy makers in implementing such integrated tender systems.

As the Dutch benchmark is established with the interviewee #10, the other interviews are used to evaluate how the APAC markets compare with the Dutch model, arguably defined as the "best in class".

Regarding policy instruments for the allocation of large renewable energy projects, the transition to a subsidy free auction system has not yet happened. Japan, Korea, Taiwan, and Australia still have policy instruments providing substantial subsidies for renewable energy project developers. The main reason is that the maturity of the market is far off the practices that can be observed in Europe. Developers and investors need confidence regarding the viability of their business cases. However, interviewees have recognized that policy makers are considering with interest how Germany and the Netherlands are developing their own policies. Further changes towards less subsidized policy instruments shall be expected in the mid-term.

Significant grid infrastructure upgrade and development will be required to accommodate large intermittent renewable electricity into the grid. Building flexibility in the system will be required but appear to be challenging. Japanese, Korean and Taiwanese grids have been designed to accommodate large amount of baseload electricity (i.e. coal, nuclear...). Unbundling of the electricity sector has just begun, and it will take years before it materializes (i.e. Korea and Japan). This may hinder easy access to the grid. Interconnections between grids is poor to inexistent (i.e. Japan, Australia...). Finally, models used for the calculation of the grid capacity are underestimating the actual grid capacity. In this context, renewable hydrogen could play a role in providing flexibility to the grid and enable the development of the large renewable energy projects.

There are no decarbonization targets for the industry. Whether in Japan, Korea or Australia, governments are still choosing industrial development over environmental protection, despite having signed the COP21 Paris agreement. Some industrial actors are considering hydrogen as an alternative energy vector. However, renewable hydrogen is not yet considered as an economically viable option.

When the targeted APAC countries signed the COP21 Paris agreement, they committed to act for limiting their impact on the climate change. One action is the development of the renewable energy sector. However, the level of development of large renewable energy projects is contrasted depending on the APAC countries. Onshore renewable energy development in Korea is limited because of the lack of space (i.e. large flat areas for solar PV). In Japan, electricity is expensive in general and offshore wind is estimated to be produced at 250 euros per megawatt. Other carbon neutral technologies such as nuclear are still on the agenda of Korea and Japan, directly competing with renewable energy. Australia still has a strong coal industry. Though renewable energy is developing, Korea and Japan will have difficulties to reach their targets. Taiwan seems more committed while the Federal Australian government has recently transferred its "Climate change function" to the Department of Industry, Science, Energy and Resources. Targets which were defined until 2020 are currently under review.

All the targeted APAC countries see an opportunity in developing a so-called hydrogen economy. Whereas the Netherlands see the development of the renewable hydrogen as an opportunity to address the climate change challenges, the APAC markets have taken an interest on hydrogen for different reasons. Korea and

Japan are first and foremostly interested in hydrogen as they want to take the leadership in technologies and applications using hydrogen for exportation purpose. Applications include mobility, power generation or fuel cells. Secondly, Korean and Japan see hydrogen as an opportunity to diversify their energy imports. Japan is currently developing shipping lines to import hydrogen from where it can be produced cheaply. Australia, which has a proven history of exporting commodities globally, sees the hydrogen as another export item and therefore, as another business opportunity. In this context, the renewable hydrogen is not the preferred option simply for costs reason.

In Korea, the carbon market is limited and does not provide any incentives for decarbonisation of the industry. In Japan, the effect of a carbon tax on oil and gas products has led to the development of a prosperous coal industry. Coal is the cheapest source of energy in Japan. In Australia, the Federal government has not committed on any target in order not to hinder the economic development.

All targeted APAC markets have near-shore petrochemical clusters with existing natural gas infrastructure. In those clusters, hundreds of thousands of tons of hydrogen are produced and consumed every year. Almost all the hydrogen production is done via SMR (steam methane reformation) or coal gasification in China. If a share of this hydrogen is exchanged as a commodity between industries within clusters, most of the hydrogen is produced and used in-situ. Billions of US dollars have been invested in those facilities and without any incentives for decarbonising, petrochemical industry actors do not seem to consider renewable hydrogen as a viable option for now.

Some of the current limitations seen in the Dutch market are also present in the APAC markets.

There is no definition for the term "renewable hydrogen" in any of the APAC markets. The renewable nature of hydrogen is not a driving factor of the development of the so-called hydrogen economy. Korea and Japan are investing in the research and development of applications running on hydrogen. In this context, the source of the hydrogen is not relevant. Cost is the driver. With high renewable energy production costs, the development of a domestically produced renewable hydrogen is not viable in Japan and Korea now. In Australia, the Australian National Hydrogen Strategy talks about "clean" hydrogen without providing a clear definition of its meaning. Hydrogen can be produced from coal gasification + CCUS (carbon capture, use and storage) but can also be produced from renewable energy. The strategy is based on cheap production of hydrogen and exportation to countries such as Japan. Some Australian states have however excellent renewable energy resource (i.e. solar PV in South Australia) and consider playing an important role in the Australian hydrogen strategy.

None of the targeted APAC countries have established policy instruments (i.e. subsidy scheme) for renewable hydrogen. In Korea, electricity produced from fuel cells is subject to feed in tariffs. But there is no requirement regarding the nature of the hydrogen, whether it is from renewable origin or not.

Whereas the Netherlands are considering the possible integration of offshore wind and renewable hydrogen in their tender system for new projects, in the APAC markets, the policy instruments for the development of renewable energy projects are totally disconnected and independent from those supporting the development of the so-called hydrogen economy.

The hydrogen strategies of Korea and Japan are putting emphasis on the mobility (i.e. Hyundai, Toyota...). However, the mobility sector does not provide the scale effect that the petrochemical industry does. This prevents the rapid reduction of the costs. Other hydrogen-based applications being developed, such as for power generation, could require larger demand for hydrogen as feedstock.

Some lobbies are very influential in Japan and Korea. Fisheries strongly oppose to large offshore wind projects. Pro-nuclear and oil & gas are lobbying against the use of hydrogen in the mobility sector due to safety risks.

5.2.6   The Concluding Remarks Based on the Interview Findings

Hydrogen, as an alternative energy vector, is high on the governmental agenda of all the targeted APAC countries. National hydrogen strategies have been published and funds are made available for research, development and pilot projects demonstrating the economically viable, technologically feasible and safe use of hydrogen.

Whereas the Netherlands are supporting the development of renewable hydrogen offtake as an enabler of the energy transition, the APAC countries discussed during the interviews see different opportunities. Korea and Japan have positioned themselves in the downstream part of the value chain. Both countries are developing applications to store and use the hydrogen, with as a strategy, to export those applications. Australia, at the other end of the value chain, is focusing on the production of cheap hydrogen for export purpose. Both ends of the value chain are critical for the success of the so-called hydrogen economy.

However, hydrogen from renewable energy source is not yet at the centre of the APAC countries hydrogen strategies. This is due to the following factors:

- Fossil-fuel and nuclear baseload power has been the backbone of the energy system of the APAC markets for decades. The transition to a renewable energy system is happening, but at a much slower pace than in Europe.
- There is no target or incentive for the industry to decarbonize.
- The expensive and still limited quantity of renewable energy capacity makes cost of the renewable hydrogen too high compared to the fossil fuel-based hydrogen.
- The policy instruments for the development of renewable energy projects protect developers and investors from the merchant risk.

Nonetheless, renewable hydrogen could find applications in the future in the APAC markets.

Hundreds of thousands of tons of hydrogen are used in various industrial processes (i.e. fertilizer manufacturers, steel manufacturers, refineries…).

Growing grid capacity and curtailment issues call for solutions in Japan, Korea and Australia. Renewable hydrogen can play a role in bringing flexibility to the system.

Countries such as Australia and New-Zealand have excellent renewable energy sources. They are developing strategies to produce renewable hydrogen for exportation purpose.

Some policy instruments could be developed to support the development of the renewable hydrogen production and offtake:

- Decarbonisation targets for the industry
- Subsidies to produce renewable hydrogen or energy (i.e. electricity, heat…) from renewable hydrogen
- Certification of the renewable hydrogen
- Co-development of the policy instruments for renewable energy projects and renewable hydrogen based on the Dutch model (i.e. integrated tender)

Renewable energy policies are at their early days in the APAC markets. There is an opportunity to shape those policies based on the successful models developed in more mature markets such as the Dutch one.

5.3 Case Studies for Determining the Potential of Renewable Hydrogen for Decarbonising the Industrial Sectors Using Fossil Fuel-Based Hydrogen within the APAC Markets

The case studies chosen for the research work are five countries of the APAC markets: Australia, China, Japan, Korea, and Taiwan.

The survey of the five case studies has for objective to provide an order of magnitude of the potential for decarbonisation by using renewable hydrogen.

The survey is based on the data analysis of the documents identified in the literature review and some other data sources used while performing the survey. The content of the survey is built using the *renewable hydrogen framework* (figure 5.4) of the present document.

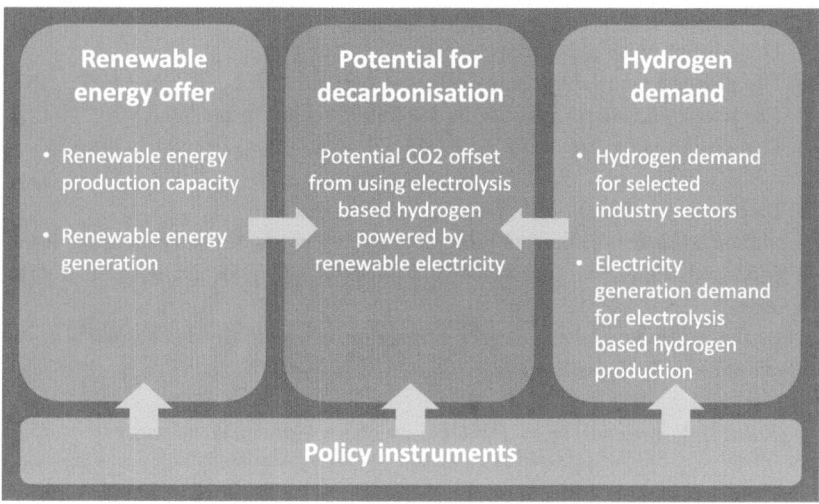

Figure 5.4 Building blocks of the survey design

Accordingly, the survey covers the status of the current renewable energy generation capacity and production. The renewable electricity generation is also estimated based on the renewable energy targets (2030) and the renewable energy technical potential. This covers the "Renewable energy offer" element of the *renewable hydrogen framework*.

Secondly, the survey covers the hydrogen demand of the five case studies resulting from the following carbon intensive industry sectors: ammonia, methanol, steel manufacturing and oil refining. This information addresses the "Hydrogen demand" element of the *renewable hydrogen framework*.

Thirdly, the survey consists on combining the data collected from the two first elements (i.e. renewable energy offer and hydrogen demand) to provide an order of magnitude of the potential for decarbonisation in the chosen industry sectors. The survey design is summarized in the figure 5.4.

The underpinning policy instruments have been identified during the literature review.

5.3.1   Determining the Renewable Energy offer

The determination of the renewable energy offer is broken down into four steps:

(1) The current renewable energy offer
(2) The planned renewable energy offer (at horizon 2030)
(3) The renewable energy offer at its maximum technical potential
(4) The scenarios for providing renewable electricity for electrolysis purpose

The current renewable energy offer
The author estimates the current renewable energy production based on the installed wind and solar PV capacity and the corresponding local renewable energy resources.

For the estimation of the yearly generation of electricity from solar PV, the following formula is used:

$$Solar\ PV\ generation = PVOUT * 365 * Solar\ PV\ capacity \qquad (5.1)$$

Where:

- Solar PV generation is electricity generation during a year in TWh y^{-1}.
- PVOUT is the specific photovoltaic power output in TWh/TW$_p$ per day. The specific photovoltaic power output data are obtained from the "Global Solar Atlas 2.0", a free, web-based application is developed and operated by the company Solargis s.r.o. on behalf of the World Bank Group, utilizing Solargis data, with funding provided by the Energy Sector Management Assistance Program (ESMAP) (Global Solar Atlas n.d.).

- Solar PV capacity is the installed solar PV capacity in TW obtained from the "Renewable capacity statistics 2019" (IRENA 2019).

For the estimation of the yearly generation of electricity from wind power, the following formula was used:

$$Wind\, generation = P_{rwt} * Number\, of\, reference\, wind\, turbines * 8760$$
(5.2)

Where:

- $Wind\, generation$ is the electricity generation during a year in TWh y^{-1}.
- P_{rwt} is power produced by one reference wind turbine in TW, as per formula (5.3).
- $Number\, of\, reference\, wind\, turbines$ is the total wind power installed capacity divided by the nameplate nominal power of the reference wind turbine, as per formula (5.4)
- 8760 in hours of operation per year (assumed).

$$P_{rwt} = \frac{\frac{1}{2} * \rho * A * c_P * v^3}{1 * 10^{12}}$$
(5.3)

Where:

- ρ is the density of air assumed to be uniformly of 1,23 kg/m^3 across all the case studies.
- A is the swept area of the reference wind turbine in m^2, as per formula (5.5).
- c_P is the unitless capacity factor of the reference wind turbine. The capacity factor is assumed to be of 0.367 for onshore wind and 0.5 for offshore wind in all study cases.
- v is the case studies specific average wind speed in m/s. The average wind speed is obtained from the "Global Wind Atlas 3.0", a free, web-based application developed, owned and operated by the Technical University of Denmark (DTU). The Global Wind Atlas 3.0 is released in partnership with the World Bank Group, utilizing data provided by Vortex, using funding provided by the Energy Sector Management Assistance Program (ESMAP) (Global Wind Atlas n.d.).

$$Number\ of\ reference\ wind\ turbines$$
$$= \frac{Total\ wind\ power\ capacity\ of\ the\ case\ study}{Nameplate\ ominal\ power\ of\ the\ reference\ wind\ turbine} \qquad (5.4)$$

Where:

- The *Total wind power capacity of the case study* is expressed in MW, including both onshore and offshore wind installed capacity, and obtained from the "Renewable capacity statistics 2019" (IRENA 2019).
- The *Nameplate nominal power of the reference wind turbine* is expressed in MW. A value of 1.16 MW is used uniformly for all the case studies as obtained from the "Renewable Energy Technologies: Cost Analysis Series, Volume 1: Power Sector, Issue 5/5, Wind Power" (IRENA 2012).

$$A = \frac{Nameplate\ nominal\ power\ of\ the\ reference\ wind\ turbine * 10^6}{SP} \qquad (5.5)$$

Where:

- The *Nameplate nominal power of the reference wind turbine* is in MW and as explained in formula (4).
- The *SP* is the specific power of the reference wind turbine expressed in W/m². A SP of 300 W/m2 is obtained from "System Value of Wind Power an Analysis of the Effects of Wind Turbine: Design Economic dispatch modelling of medium-term system implications of advanced wind power technologies" (Dalla Riva 2016) and used uniformly across the case studies.

For validation purpose, the author compares the actual solar PV and wind power generation figures reported in the grey and the scientific literature. As shown in table 5.3, both solar PV and wind yearly electricity generation data obtained from the calculations are in the scale of magnitude of the data found in the relevant literature. However, there are differences due to the following variables and factors (Lacerda and van den Bergh 2016):

Table 5.3 Current renewable electricity generation offer

Current		China (CNREC 2019; CNREC 2019; Yang et al. 2017)	Japan (IEA 2018b)	Korea (IEA 2018c)	Taiwan (Salt 2018)
Current onshore wind capacity in GW	6	180	4	1	1
Current offshore wind capacity in GW	-	5	0	0	0
Current onshore wind generation (calculated) in TWh yr^{-1}	20	845	11	3	5
Current offshore wind generation (calculated) in TWh yr^{-1}	-	29	0	0	0
Current wind generation (calculated) in TWh yr^{-1}	20	875	11	3	6
Current wind generation (from literature) in TWh yr^{-1}	12	366	6	2	4
Current solar PV capacity in GW	10	175	56	8	3

(continued)

- The capacity factors vary in function of the solar PV or wind turbine design and the location of the renewable energy project,
- The system flexibility varies in function of the network infrastructure and management, the portfolio diversity, the energy storage, and the demand site management,

Table 5.3 (continued)

Current		China (CNREC 2019; CNREC 2019; Yang et al. 2017)	Japan (IEA 2018b)	Korea (IEA 2018c)	Taiwan (Salt 2018)
Current solar PV generation (calculated) in TWh yr^{-1}	18	257	69	11	3
Current solar PV generation (from literature) in TWh yr^{-1}	6	178	52	5	10
Current total RE generation (calculated) in TWh yr^{-1}	37	1.131	80	14	9
Current total RE generation (from literature) in TWh yr^{-1}	18	544	58	7	13

- The market integration varies in function of the balance of dispatch, the reduced time of response, the local marginal pricing, the curtailment control and the cross-border trade.

Though some of the factors and variables have been assumed and implemented uniformly across the case studies (i.e. capacity factor, nameplate nominal power of the reference wind turbine), most of them are not included in the author's calculations.

For calculating the current renewable electricity offer, the author uses the generation data found in the literature. The most conservative renewable electricity generation figure is chosen in case multiple sources are providing different data. For calculating the planned and potential renewable electricity generation data, the author uses the formulae described in the section with a correction factor. The correction factor is specific to each case studies. It is obtained by dividing the current renewable electricity generation from the literature by the calculated current renewable electricity generation.

The planned renewable energy offer (2030)

Table 5.4 Planned renewable energy offer at 2030

2030	Australia (Australian Govern-ment n.d.)	China (CNREC 2019)	Japan (Institute for Sustain-able Energy Policies 2019)	Korea (Lee 2019)	Taiwan (L. Qiao 2019; R. Wang 2017)
Planned onshore wind capacity in GW	–	1.023	27	10	1
Planned offshore wind capacity in GW	–	27	10	0	16
Planned onshore wind generation (calculated) in TWh yr^{-1}	–	4.802	84	25	9
Planned offshore wind generation (calculated) in TWh yr^{-1}	–	171	42	1	166
Planned wind generation (calculated) in TWh yr^{-1}	–	4.973	126	27	176
Planned wind generation (correction factor) in TWh yr^{-1}	–	2.081	67	13	118

(continued)

As identified in the literature review, each case study has defined renewable energy targets. To perform the assessment of the decarbonisation potential, the author uses the targets at year 2030. Australia has used a renewable electricity generation target of 33 terawatts hour by 2030. The other study cases have defined renewable

Table 5.4 (continued)

2030	Australia (Australian Govern-ment n.d.)	China (CNREC 2019)	Japan (Institute for Sustain-able Energy Policies 2019)	Korea (Lee 2019)	Taiwan (L. Qiao 2019; R. Wang 2017)
Planned solar PV capacity in GW	–	1.195	100	19	20
Planned solar PV generation (calculated) in TWh yr^{-1}	–	1.753	124	26	25
Planned solar PV generation (correction factor) in TWh yr^{-1}	–	1.215	93	13	25
Planned total RE generation (calculated) in TWh yr^{-1}	33	6.727	250	53	201
Planned total RE generation (correction factor) in TWh yr^{-1}	33	3.297	160	25	144

energy capacity targets. Taiwan targets are set only until 2025 and were used as such for the 2030 calculations.

The planned renewable energy capacity and generation in the APAC markets falling in the research scope are provided in the table 5.4.

The case studies potential for renewable energy offer

The renewable energy technical potential is the theoretical maximum capacity that can be installed with the current technology. This data is identified from other sources than from the literature review. Based on the maximum technical potential of each study cases, the author calculates the corresponding generation using the same correction factors determined in the table 5.3.

The technical potential for renewable energy capacity and generation in the APAC markets falling in the research scope are provided in the table 5.5.

Table 5.5 Technical potential for renewable energy capacity and generation

Technical potential	Australia (ISF 2016)	China (World Bank Group 2020; IRENA 2014)	Japan (GENI 2012)	Korea (Hong et al. 2019)	Taiwan (Lu 2016)
Technical potential onshore wind capacity in GW	880	1.950	283	64	4
Technical potential offshore wind capacity in GW	660	2.982	1.500	33	74
Technical potential onshore wind generation (calculated) in TWh yr^{-1}	2.981	9.152	878	166	31
Technical potential offshore wind generation (calculated) in TWh yr^{-1}	3.046	19.067	6.342	118	788
Technical potential wind generation (calculated) in TWh yr^{-1}	6.026	28.219	7.221	285	820

(continued)

Table 5.5 (continued)

Technical potential	Australia (ISF 2016)	China (World Bank Group 2020; IRENA 2014)	Japan (GENI 2012)	Korea (Hong et al. 2019)	Taiwan (Lu 2016)
Technical potential wind generation (correction factor) in TWh yr^{-1}	3.744	11.809	3.824	136	553
Technical potential solar PV capacity in GW	24.100	2.200	150	3.183	74
Technical potential solar PV generation (calculated) in TWh yr^{-1}	43.543	3.228	186	4.415	103
Technical potential solar PV generation (correction factor) in TWh yr-1	15.225	2.237	140	2.136	103

(continued)

The scenarios for providing renewable electricity for water electrolysis purpose
Three scenarios are considered to determine the quantity of renewable electricity available to produce hydrogen via water electrolysis.

The scenario (1) corresponds to the situation where all the renewable electricity produced is dedicated to the water electrolysis to produce hydrogen.

Table 5.5   (continued)

Technical potential	Australia (ISF 2016)	China (World Bank Group 2020; IRENA 2014)	Japan (GENI 2012)	Korea (Hong et al. 2019)	Taiwan (Lu 2016)
Technical potential total RE generation (calculated) in TWh yr^{-1}	49.569	31.447	7.406	4.699	923
Technical potential total RE generation (correction factor) in TWh yr^{-1}	18.969	14.046	3.964	2.273	656

The scenario (2) corresponds to the situation where only the curtailed renewable electricity is used to produce hydrogen. To conduct this scenario, the author uses the curtailment ratio obtained from the relevant literature.

The scenario (3) establishes that in a liberalised energy market highly penetrated by renewables, the production of renewable hydrogen by water electrolysis is profitable 2550 hours a year (van Leeuwen and Mulder 2018).

5.3.2   Determining the Hydrogen Demand

The research work focuses on the hydrogen demand for carbon intensive industry sectors such as the petrochemical feedstock production, the oil refining, and the steel manufacturing.

The hydrogen demand assessment is based on the production data of the case studies. The production of ammonia [Table 5.6] and methanol [Table 5.7] data are used for the petrochemical feedstock production.

Table 5.6 Methanol production in Mt yr^{-1}

Country	Ammonia production	Comments	References
Japan	–	No production	(MMSA 2012)
China	56,10		(International ASA 2018)
Taiwan	–	No production	(FAO 2002)
Australia	1,30		(US Geological Survey 2019)
Korea	–	No production	(MMSA 2012)

Table 5.7 Ammonia production in Mt yr^{-1}

Country	Methanol production	Comments	References
Japan	–	No production	(MMSA 2012)
China	65,70		(Xu et al. 2017)
Taiwan	–	No production	(MMSA 2012)
Australia	0,08		(MMSA 2012)
Korea	–	No production	(MMSA 2012)

The oil processed data [Table 5.8] are used for the oil refining.

The pig iron production data from the BF-BOF process [Table 5.9] and DRI-EAF processes data are used for the steel manufacturing.

A conversion factor is required to estimate the hydrogen demand based on the production data.

Depending on the configuration of the plant and the production process, the production of 1 ton of ammonia requires 178,18–182,44 kg of hydrogen (Matijašević and Petrić 2016). For calculation purpose, the author uses the requirement of 0,18 ton of hydrogen to produce 1 ton of ammonia.

The production of 1 ton of methanol requires 126,45–142,26 kg of hydrogen (Matijašević and Petrić 2016). For calculation purpose, the author uses the requirement of 0,135 ton of hydrogen to produce 1 ton of methanol.

Table 5.8 Oil processed in refineries in Mt yr^{-1}

Country	Oil processed in refineries	References
Japan	153,21	(UNSD — Energy Statistics n.d.)
China	632,73	(UNSD — Energy Statistics n.d.)
Taiwan	66,68	(MOEA 2019)
Australia	22,21	(UNSD — Energy Statistics n.d.)
Korea	155,57	(UNSD — Energy Statistics n.d.)

Table 5.9 Pig iron production from BF-BOF process in Mt yr^{-1}

Country	Option 1: BF-BOF steel manufacturing (World Steel Association 2019)	Recycled metal share in %	Option 2:	
			Iron ore DRI-EAF steel manufacturing	Recycled metal DRI-EAF steel manufacturing
Japan	77,33	32% (Åhman et al. 2018)	52,58	24,74
China	771,05	11% (Dolci 2018)	686,24	84,82
Korea	3,88	39% (Dolci 2018)	2,37	1,51
Taiwan	47,12	30% (assumed)	32,99	14,14
Australia	14,84	30% (assumed)	10,39	4,45

Concerning oil refining, processing 1 ton of crude oil requires about 50 Nm^3 or 0,0041 ton of hydrogen (Matijašević and Petrić 2016). Other sources indicate that 0,0048 ton of hydrogen is required to process 1 ton of crude oil (Dolci 2018). For calculation purpose, the author uses the hydrogen requirement of 0,0045 ton of hydrogen to process 1 ton of crude oil.

The BF-BOF steel manufacturing process does not require hydrogen as the reduction of the iron ore is done by the carbon contained in the coke. The reduction process can be decarbonised by injecting hydrogen together with the coke. The injection of 27 kg of hydrogen per ton of pig iron produced can reduce the CO_2 emissions by 20% (Dolci 2018). The calculations for the hydrogen demand from the steel manufacturing considers two options. The option (a) corresponds to the carbon abated BF-BOF process by injecting 0,0027 t of hydrogen per ton of iron pig produced. The option (b) considers the substitution of the BF-BOF process by the carbon free DRI-EAF process where the reduction of iron ore is done entirely by hydrogen (Vogl, Åhman, and Nilsson 2018). The DRI process is also used for the manufacturing of steel from recycled metal. Therefore, the hydrogen demand from the DRI process shall distinguish the manufacturing of virgin steel from iron ore path from the recycled metal path which is a 50%/50% mix of iron ore and recycled metal. The DRI process requires 0,0051 ton of hydrogen per ton of steel produced from iron ore and 0,0025 ton of hydrogen per ton of steel produced from recycled metal (Vogl, Åhman, and Nilsson 2018). The hydrogen demand of the option (b) is calculated assuming that all the steel manufactured via BF-BOF is now manufactured via DRI-EAF process considering the recycled metal share when applicable. As of today, none of the study cases are manufacturing steel with the DRI process at industrial scale (World Steel Association 2019). The calculation of the hydrogen demand from the DRI-EAF steel manufacturing process is done for estimation purpose, should the BF-BOF steel manufacturing process be completely decarbonised. The share of recycled metal manufacturing is increasing globally (Åhman et al. 2018). It is however still low for the case studies with 32% in Japan, 39% in Korea and only 11% in China compared to 54% in the EU and 72% in the US. The author assumes a 30% share for Australia and Taiwan.

Once the production data and conversion factors are defined, the author calculates the hydrogen demand as shown in table 5.10.

As the hydrogen demand is calculated, the corresponding electricity demand to produce hydrogen from water electrolysis can be determined. The PEMEC technology is becoming the leading technology due to its superior characteristics for intermittent operations, consistent with the renewable electricity energy supply. 59.94 kWh of electricity are required to produce 1 kg of hydrogen (Schmidt et al. 2017). Hydrogenics®, a manufacturer of PEMEC water electrolyser, indicates that + /– 55 kWh of electricity are required to produce 1 kg of hydrogen (Hydrogenics 2018). The author uses 55 kWh for the calculations displayed in the table 5.11.

Table 5.10   Hydrogen demand per production process in Mt y^{-1}

Country	Japan	China	Korea	Taiwan	Australia	Total per process
Ammonia production	-	10,10	-	–	0,23	**10,33**
Methanol production	–	8,87	–	–	0,01	**8,88**
Oil refining	0,68	2,82	0,69	0,30	0,10	**4,59**
Carbon abated BF-OF steel manufacturing process (option 1)	2,09	20,82	1,27	0,40	0,10	**24,68**
DRI-EAF steel manufacturing process (incl. recycled share) (option 2)	2,85	37,23	1,56	0,56	0,15	**42,36**
Total per country with steel manufacturing option 1	**2,77**	**42,60**	**1,97**	**0,70**	**0,45**	**48,49**
Total per country with steel manufacturing option 2	**3,54**	**59,02**	**2,25**	**0,86**	**0,49**	**66,16**

5.3.3   Determining the Potential for Decarbonisation

The determination of the decarbonisation potential is done by calculating the offset of CO_2 emissions if the hydrogen used in the industrial process is produced from water electrolysis powered by renewable electricity.

As identified in the literature review, the CO_2 emissions offset depends on the fossil fuel-based hydrogen production processes. The SMR process releases 12 kg of CO_2 per kg of hydrogen produced while the coal gasification releases 27 kg of CO_2 per g of hydrogen produced (Simons and Bauer 2011).

In the oil refining sector, hydrogen is also produced as a by-product of steam naphtha reforming (catalytic reforming) process. The CO_2 emissions of steam naphtha reforming (catalytic reforming) is of 10,5 kg per kg of hydrogen produced (EPA 2008). A portion of the hydrogen used in oil refining is also purchased. This merchant hydrogen for oil refining is assumed to be produced from SMR in the calculations.

Table 5.11 Renewable electricity demand for water electrolysis hydrogen production in TWh y^{-1}

Country	Japan	China	Korea	Taiwan	Australia	Total per process
Ammonia production	–	560,44	–	–	12,99	**573,43**
Methanol production	–	492,26	–	–	0,60	**492,86**
Oil refining	37,87	156,40	38,46	16,48	5,49	**254,70**
Carbon abated BF-OF steel manufacturing process (option 1)	115,88	1 155,42	70,62	22,24	5,82	**1 369,97**
DRI-EAF steel manufacturing process (incl. recycled share) (option 2)	158,34	2 066,38	86,56	31,28	8,18	**2 350,74**
Total per country with steel manufacturing option 1	**153,75**	**2 364,52**	**109,07**	**38,72**	**24,89**	**2 690,96**
Total per country with steel manufacturing option 2	**196,21**	**3 275,48**	**125,01**	**47,76**	**27,26**	**3 671,72**

The share of fossil fuel-based hydrogen production processes depends from the case studies and from the manufacturing process it is intended for as shown in table 5.12.

As identified in the literature review, there are different paths for manufacturing steel with different CO_2 emissions intensities. The most common process to manufacture steel is the BF-BOF process for which 1,870 tons of CO_2 are emitted during the production of 1 ton of steel (Vogl, Åhman, and Nilsson 2018). By adding 0.027 kg of hydrogen in the blast furnace, it is possible to reduce the CO_2 emissions by 20% (Dolci 2018). This carbon abatement route will be followed-up as an option (a) for steel manufacturing decarbonisation.

Table 5.12   Distribution of fossil fuel-based hydrogen production processes per industry sector

Country	Industrial sectors	Hydrogen production process				
		SMR	Catalytic reforming	Merchant	Coal gasification	References
Japan	Oil refining	39%	33%	25%	3%	(IEA 2019)
China	Ammonia	22%	0%	0%	78%	(Saygın et al. 2009)
	Methanol	0%	0%	0%	100%	(Saygın et al. 2009)
	Oil refining	28%	42%	17%	13%	(IEA 2019)
Korea	Oil refining	39%	33%	25%	3%	(IEA 2019)
Taiwan	Oil refining	39%	33%	25%	3%	(IEA 2019)
Australia	Ammonia	50%	0%	0%	50%	Assumed
	Methanol	50%	0%	0%	50%	Assumed
	Oil refining	39%	33%	25%	3%	(IEA 2019)

The DRI-EAF route is the second most common process for steel manufacturing. It uses methane to reduce the iron ore. The DRI phase releases 0,6 tons of CO_2 per ton of virgin steel produced. The EAF phase is powered by electricity and due to the electricity generation mix, CO_2 emissions are estimated to be of 0,3 tons of CO_2 per ton of steel produced (ETC/RMI 2019; De Pee et al. 2018). Currently, none of the case studies are using DRI-EAF at industrial scale for steel manufacturing. The option (b) for steel manufacturing decarbonisation consists on substituting the steel manufacturing from the BF-BOF process by the DRI-EAF process. In the option (b) for steel manufacturing decarbonisation, there is no CO_2 emissions. The methane used in the DRI process is replaced by hydrogen as detailed above. The EAF phase does not require hydrogen but only electricity. For the EAF phase, 0,753 MWh of electricity is required per ton of virgin steel produced while 0,667 MWh of electricity is required per ton of mix virgin and recycled steel produced (Vogl, Åhman, and Nilsson 2018).

As the potential for carbon offset and the corresponding amount of renewable electricity required to produce renewable hydrogen have been calculated, it is

Table 5.13   Potential CO_2 offset (from use of renewable hydrogen) in Mt y^{-1}

Country	Japan	China	Korea	Taiwan	Australia	Total per process
Ammonia production	–	239,32	–	–	4,56	**243,89**
Methanol production	–	239,48	–	–	0,21	**239,69**
Oil refining	8,16	37,54	8,28	3,55	1,18	**58,71**
Carbon abated BF-OF steel manufacturing process (option 1)	30,77	306,78	18,75	5,90	1,54	**363,75**
DRI-EAF steel manufacturing process (incl. recycled share) (option 2)	153,83	1 533,91	93,75	29,52	7,72	**1 818,73**
Total per country with steel manufacturing option 1	**38,92**	**823,12**	**27,03**	**9,46**	**7,50**	**906,03**
Total per country with steel manufacturing option 2	**161,99**	**2 050,24**	**102,03**	**33,07**	**13,68**	**2 361,02**

5.1

possible to estimate how much the renewable electricity generation of the study cases can fulfil the renewable hydrogen demand. The order of magnitude for decarbonisation potential for a given study case is obtained from the following formula:

$$Decarbonisation\ potential = \frac{Available\ RE_{Generation}}{Required\,RE_{Generation}\ for\ H_{2RE}\ production} \quad (6)$$

Where:

- The decarbonisation potential is expressed in percentage.

- The *Available $RE_{Generation}$* is the renewable electricity generation in TWh available in the study case. The *Available $RE_{Generation}$* is calculated following three scenarios defined end of section 5.3.1.
- The *Required $RE_{Generation}$ for H_{2RE} production* is the electricity generation in TWh calculated to produce the renewable hydrogen required for decarbonising the industry sectors covered by the research work.

The decarbonisation potential is assessed against the study cases renewable energy targets and their respective renewable energy potential.

When estimating the decarbonisation potential against the 2030 targets and the technical potential renewable energy capacities, the hydrogen demand is adjusted to the projected demand in 2030 [Table 5.14]. The steel manufacturing options (a) and (b) are however not amended. A gradual replacement of the BF-BOF steel manufacturing process by the DRI-EAF is expected over the coming decades. The option (b) corresponds to a scenario where all steel manufacturing is done using the DRI-EAF process.

Table 5.14 Hydrogen demand projected evolution by 2030 [3]

Process	Hydrogen demand increase by 2030
Oil refining	7%
Ammonia	31%
Methanol	31%
Steel manufacturing	100% (shift to DRI process)

Summary of chapter 5

The chapter 5 has answered the secondary question:

What is the hydrogen demand of industry sectors using fossil fuel-based hydrogen as feedstock?

The chapter 5 has supplemented the chapter 4 by answering the secondary question from a quantitative point of view.

The survey of the study cases has allowed determining the quantity of hydrogen required in the production processes of ammonia, methanol, steel, and oil refining. It is then possible to estimate the amount of renewable electricity required for the decarbonization of those production processes with renewable hydrogen.

The literature review and the semi-structured interviews findings have clarified the goals and implications of the APAC markets renewable energy and hydrogen policies. The design and the implementation of the respective policies are not linked. Moreover, the APAC hydrogen policy frameworks reviewed in the chapter 5 are often technology neutral, without specifically referring to the renewable hydrogen.

The quantitative findings of the chapter 5 are used to determine the decarbonization potential as discussed in the chapter 6.

Results

The results of the survey are displayed using the template "Country decarbonisation potential assessment" available in appendix 18.

On the left end side of the template, the renewable energy offer is presented according to the current, planned at 2030 horizon, and technical potential capacities for solar PV and wind energy. The amount of renewable electricity available for producing hydrogen through water electrolysis is displayed following the three scenarios depicted in section 5.3.1 of this paper: all renewable electricity is used, only the curtailed portion is used, and 2.550 hours of renewable electricity production is used.

On the right end side of the template, the hydrogen demand is presented for each industrial process falling under the scope of this paper. As the hydrogen demand is determined, the corresponding CO_2 emissions when hydrogen is produced from fossil fuels are displayed. Similarly, the electricity required to produce the hydrogen through water electrolysis is shown. Current and 2030 projected hydrogen demand figures are provided.

In all cases the scenarios involving the steel manufacturing option (a) would only lead to a partial decarbonisation as some coke would steel be used for the reduction of the iron ore.

For all study cases, the decarbonisation potential is only estimated based on the large solar PV and wind power renewable capacity. Large hydropower electricity generation capacity is not considered.

The decarbonisation potential is given in the central part of the template. It makes the link between the renewable energy offer and the hydrogen demand.

Electronic supplementary material The online version of this chapter (https://doi.org/10.1007/978-3-658-32642-5_6) contains supplementary material, which is available to authorized users.

M. Jaunatre, *Renewable Hydrogen*, Business Analytics,
https://doi.org/10.1007/978-3-658-32642-5_6

Considering that the renewable energy offer is representative of the renewable energy policies in force, the decarbonisation potential, expressed in percentage, provides an indication on how much those policies can impact of the decarbonisation of industry sectors using fossil fuel-based hydrogen.

The following section provides the results of the decarbonisation potential assessment, case study by case study. At the end of the section, he results are consolidated to provide an APAC and regional wide overview of the research topic.

6.1   The Results of the Study Case Australia

6.1.1   The Current Decarbonisation Potential of Australia

The appendix 19 shows the decarbonisation potential under the current renewable energy policy regime in Australia. The steel manufacturing option (b) is not a viable scenario as that would mean that all the steel manufacturing currently happening in Australia would be changed to the DRI-EAF route.

The results show that if all the renewable electricity produced in Australia would be used to produce renewable hydrogen as per scenario (1), their industries falling under the research scope could be decarbonised to 74%, offsetting 5,55 Mt of CO_2.

In the more realistic scenario (2), the decarbonisation potential would be of 4%, with 0,33 Mt of CO_2 offset.

In the scenario (3), using a power-to-gas system for a duration of 2.550 hours yr^{-1} would enable a 21% decarbonisation, offsetting 1,61 Mt of CO_2.

6.1.2   The 2030 Decarbonisation Potential of Australia

The appendix 19 shows the decarbonisation potential with the 2030 targeted renewable energy capacities set under the current renewable energy policies in Australia.

The scenario (1) with steel manufacturing option (b) shows that with the current 2030 target of 33 gigawatts hour renewable electricity production, Australia would be able to entirely decarbonise their industries falling under the research scope. There would even be a slight excess generation. The CO_2 offset would be of 15,24 Mt.

Using only the curtailed renewable electricity production according to scenario (2), the decarbonisation potential would range between 6–7% depending on the steel manufacturing process chosen. The CO_2 offset could be up to 0,95 Mt.

Following the scenario (3), a profitable use of a power-to-gas system would enable a decarbonisation potential of around 30%, corresponding to a CO2 offset ranging between 2,94–4,58 Mt.

6.1.3   The Decarbonisation Potential of Australia Using its Technical Potential of Renewable Energy

The appendix 19 shows the huge renewable energy potential of Australia and its renewable hydrogen exportation capacity. In all scenarios, Australia can cover its needs, which represents a fraction of its massive potential for renewable hydrogen, the rest being available for exportation.

6.1.4   The concluding remarks regarding the impact of renewable energy policies on Australia industry decarbonisation

Taking the assumption that the level of renewable energy capacity currently under operation in Australia is the result of its renewable energy policies, the potential for decarbonisation of the industry sectors falling under the research scope is low, ranging from 1–5% with the scenarios (2) and (3). The scenario (1) is not realistic as the renewable electricity production cannot be entirely devoted to the production of renewable hydrogen.

The review of the renewable energy policies at the horizon 2030 shows that Australia is committed to support the development of up to 33 gigawatts hour of renewable electricity. With this renewable electricity generation capacity, the decarbonisation potential for Australia own industry sectors using fossil fuel-based hydrogen would range between 6–32% for the scenarios (2) and (3).

However, this amount of renewable energy capacity would not be enough for Australia to be a global leader in the exportation of hydrogen. To reach the goal defined in the Australia National Hydrogen Strategy, more renewable energy capacity is required. This additional renewable energy generation will have to be developed outside the current renewable energy support scheme if the latter remains unchanged.

Australia has immense renewable electricity generation capacity with an estimated 18.969 terawatts hour at technical potential. With 340 Mt yr^{-1} of potential renewable hydrogen production if all the renewable electricity produced at technical potential would be used for electrolysis purpose, this would cover Australia's hydrogen needs and fulfil its ambition of global leadership in hydrogen exportation.

6.2   The Results of the Study Case China

6.2.1   The Current Decarbonisation Potential of China

Due to its large consumption of hydrogen, the decarbonisation potential of China using the current renewable electricity production is limited, as shown in the appendix 20. Like the Australia study case, the steel manufacturing option (b) is not a viable scenario as that would mean that all the steel manufacturing currently happening in China would be changed to the DRI-EAF route.

As per scenario (1), even if all the renewable electricity generated in China is utilized to produce renewable hydrogen, the decarbonisation potential of the industry sectors under the research scope would be of only 23% for the steel manufacturing option (a).

Following the scenario (2), using the curtailed renewable electricity would enable a decarbonisation potential of about 2% for the steel manufacturing option (a) with a 19,20 Mt of CO_2 offset.

With the scenario (3) and the steel manufacturing option (a), the decarbonisation potential would be of 7% with a 54,92 Mt CO_2 offset.

6.2.2   The 2030 Decarbonisation Potential of China

The appendix 20 shows the decarbonisation potential of China with the targets for renewable electricity generation capacity at 2030. According to the scenario (1) and depending on how much the steel manufacturing industry has shifted from the option (a) to the option (b), China could be able to cover its hydrogen demand if all the renewable electricity produced would be used for water electrolysis purpose. However, a full decarbonisation would not be possible as only a 92% of decarbonisation potential could be reach with a full conversion to steel manufacturing process option (b).

With the scenario (2) using the curtailed renewable electricity, the decarbonisation potential could range between 9–12% and corresponding to a CO_2 offset ranging between 121,73–205,71 Mt yr^{-1}.

According to scenario (3), the utilization of a power-to-gas system for a minimum economically viable period, would enable a decarbonisation potential ranging from 26–35% with a CO_2 offset ranging between 344,70–582,50 Mt yr^{-1}.

6.2.3   The Decarbonisation Potential of China Using its Technical Potential of Renewable Energy

The appendix 20 shows that China solar PV and wind power technical potential would enable a complete decarbonisation of the industry sectors falling under the research scope for scenarios (1) and (3). In the scenario (3), if China would have shifted the manufacturing process to a full DRI-EAF route as per option (b), this would enable a CO_2 offset of 2.201,30 Mt yr^{-1}.

Only using the curtailed renewable electricity generation according to scenario (2) would enable a decarbonisation potential ranging between 43–57% for a CO_2 offset ranging between 550,66–945,77 Mt yr^{-1}.

6.2.4   The Concluding Remarks Regarding the Impact of Renewable Energy Policies on China Industry Decarbonisation

China renewable electricity generation targets at 2030 is estimated to be of 3.297 terawatts hour. This is a hundred-fold the Australia 2030 renewable electricity generation targets. However, given the scale of the hydrogen demand (65,09 Mt yr^{-1}), this will only allow a partial decarbonisation potential of the industry sectors falling under the research scope.

The production of renewable electricity to China technical potential and the use of power-to-gas systems for a minimum economical duration would enable a full decarbonisation of the industry sectors mentioned above.

6.3   The Results of the Study Case Japan

6.3.1   The Current Decarbonisation Potential of Japan

According to the results shown in appendix 21, the renewable energy offer of Japan is limited. The scenario (1) indicates that if Japan would use the entire renewable electricity production, the decarbonisation potential would be of 38%, offsetting 14,63 Mt yr^{-1} of CO_2 (table). The use of the curtailed renewable electricity according the scenario (2) would only allow for a decarbonisation potential of 3%. Finally, the scenario (3) would enable a decarbonisation potential of 11%, with a CO_2 offset of 4,24 Mt yr^{-1}.

Like the other study cases, the option (b) is not viable for assessing the current decarbonisation potential as this would entail that all the steel is manufactured via the DRI-EAF route.

6.3.2   The 2030 Decarbonisation Potential of Japan

The appendix 21 shows that, following the unlikely scenario (1) with steel manufacturing option (b), Japan could achieve an 80% decarbonisation potential with a CO_2 offset of 130,82 Mt yr^{-1}. The more realistic scenario (2) would allow for a 6–8% decarbonisation potential with a 3,23–10,47 Mt yr^{-1} CO_2 offset. Finally, the scenario (3) consisting on using a power-to-gas system for a minimum period of 2.550 h yr^{-1} would enable a decarbonisation potential of 23–30% with a CO_2 offset of 11,72–37,94 Mt yr^{-1}.

6.3.3   The Decarbonisation Potential of Japan Using its Technical Potential of Renewable Energy

The results of appendix 21 show that Japan renewable energy technical potential could allow for renewable hydrogen self-sufficiency and potential full decarbonisation of the industry sectors falling in the research scope. If Japan steel manufacturing would be converted to the DRI-EAF route corresponding to the steel manufacturing option (b), the CO_2 offset could be of 162,56 Mt yr^{-1}.

6.3.4   The Concluding Remarks Regarding the Impact of Renewable Energy Policies on Japan Industry Decarbonisation

With the current renewable energy 2030 targets, the decarbonisation potential of the industry sectors falling under the research scope is limited.

However, Japan has the renewable energy technical potential to be self-sufficient in renewable hydrogen. The estimated 3,58 Mt yr^{-1} of renewable hydrogen demand at 2030 could be covered in all scenarios. The scenario (3) indicates that Japan could use economically viable power-to-gas systems to cover the renewable hydrogen demand for the ammonia, methanol, oil refineries and steel manufacturing sectors and still having up to 17 Mt yr^{-1} of renewable hydrogen available for other applications.

6.4   The Results of the Study case South-Korea

6.4.1   The Current Decarbonisation Potential of South-Korea

According to the results shown in appendix 22, the renewable energy offer of South-Korea is limited. According to scenario (1) with steel manufacturing option (a), all the renewable electricity of South-Korea would only allow for a 6% decarbonisation potential with 1,72 Mt yr^{-1} CO_2 offset.

The scenarios (2) and (3) are providing negligible decarbonisation potential.

Like the other study cases, the option (b) is not viable for assessing the current decarbonisation potential as this would entail that all the steel is manufactured via the DRI-EAF route.

6.4.2   The 2030 Decarbonisation Potential of South-Korea

The appendix 22 shows that the planned renewable energy development targets for 2030 are not considering the production of renewable hydrogen. While South-Korea plans to use 30% of renewable hydrogen in 2040 (Arabzadeh, Pilpola, and Lund 2019), using the entirety of the renewable energy generation would enable a decarbonisation of 20–23% with a limited CO_2 offset of 6,27–20,38 Mt yr^{-1}.

Like with the current renewable energy offer, the 2030 renewable energy targets with the scenarios (2) and (3) would allow for a very limited decarbonisation potential of 2–7% with a CO_2 offset of 0,5–5,91 Mt yr^{-1}.

6.4.3   The Decarbonisation Potential of South-Korea Using its Technical Potential of Renewable Energy

The results visible in the appendix 22 show a similar situation like Japan. The South-Korea renewable energy technical potential could allow for renewable hydrogen self-sufficiency and potential full decarbonisation of the industry sectors falling in the research scope. If South-Korea steel manufacturing would be converted to the DRI-EAF route corresponding to the steel manufacturing option (b), the CO_2 offset could be of 102,61 Mt yr^{-1}.

6.4.4   The Concluding Remarks Regarding the Impact of Renewable Energy Policies on South-Korea Industry Decarbonisation

The current renewable energy 2030 targets of South-Korea offer a limited potential for decarbonisation. This is not aligned with the goal defined in the Hydrogen Economy Roadmap of Korea, where 30% of the hydrogen used in 2040 should be produced from water electrolysis. The current estimation shows that with the entirety of the renewable electricity generated with the renewable energy capacity targeted at 2030, 0,46 Mt yr^{-1} of renewable hydrogen could be produced. This represents only 20% of the hydrogen demand estimated for the industry sectors falling into the research scope, which does not account for the needs of the mobility, electricity generation and heat sectors. This implies that renewable hydrogen would need to be imported if South-Korea want to reach its CO_2-free hydrogen share goals.

Like Japan, South-Korea has the renewable energy technical potential to be self-sufficient in renewable hydrogen. The estimated 2,30 Mt yr^{-1} of renewable hydrogen demand at 2030 could be covered in all scenarios. The scenario (3) indicates that South-Korea could use economically viable power-to-gas systems to cover the renewable hydrogen demand for the ammonia, methanol, oil refineries and steel manufacturing sectors and still having up to 10 Mt yr^{-1} of renewable hydrogen available for other applications.

6.5   The Results of the Study Case Taiwan

6.5.1   The Current Decarbonisation Potential of Taiwan

According to the results shown in appendix 23, the current renewable energy offer would not provide a significant decarbonisation potential. With the scenario (1) and steel manufacturing option (a), the decarbonisation potential from the entirety of the renewable electricity generation would be of 34% with a CO_2 offset of 3,26 Mt yr^{-1}.

With the scenario (2), the use of the curtailed electricity would enable a 2–3% decarbonisation potential with a 0,26–0,74 Mt yr^{-1} CO_2 offset.

The economically viable use of power-to-gas systems as per scenario (3) would allow for an 8–10% decarbonisation potential with a 0,95–2,68 Mt yr^{-1} CO_2 offset.

Like the other study cases, the option (b) is not viable for assessing the current decarbonisation potential as this would entail that all the steel is manufactured via the DRI-EAF route.

6.5.2   The 2025 Decarbonisation Potential of Taiwan

Taiwan has not defined renewable energy targets beyond 2025. However, according to the results obtained in the appendix 23, the planned renewable energy development targets for 2025 would allow for an excess of decarbonisation potential with the scenarios (1) and (3).

The most realistic scenario (3) with steel manufacturing option (b) indicates that a deep decarbonisation of the industry sectors falling in the research scope is possible. In this scenario, the economically viable use of power-to-gas systems could achieve an 85% decarbonisation potential, with a CO_2 offset of 28,37 Mt yr^{-1}.

With the scenario (2), the decarbonisation potential would be of 23–29% with a possible CO_2 offset of 2,8–7,83 Mt yr^{-1}.

6.5.3   The Decarbonisation Potential of Taiwan Using its Technical Potential of Renewable Energy

Like Japan and South-Korea, the results of appendix 23 show that Taiwan renewable energy technical potential could allow for renewable hydrogen self-sufficiency and potential full decarbonisation of the industry sectors falling in the research scope. If Taiwan steel manufacturing would be converted to the DRI-EAF route corresponding to the steel manufacturing option (b), the CO_2 offset could be of 33,32 Mt yr^{-1}.

6.5.4 The Concluding Remarks Regarding the Impact of Renewable Energy Policies on Taiwan Industry Decarbonisation

With the 2025 renewable energy targets, Taiwan is the only country from the APAC markets that could cover almost the entirety of its 2030 hydrogen demand.

The renewable electricity generation at full renewable energy technical capacity is considerably smaller than for the other APAC markets. However, Taiwan would only need to use a fraction of it to cover its hydrogen needs at 2030. This could be achieved by the profitable utilization of power-to-gas systems. With the conversion of the steel manufacturing process to the DRI-EAF pathway, Taiwan could achieve the full decarbonisation of the industry sectors under the research scope.

Summary of chapter 6

The chapter 6 has answered the secondary question:

- What is the potential of renewable hydrogen for decarbonizing the industry sectors using fossil fuel-based hydrogen as feedstock?

The chapter 6 has provided an estimate of the decarbonization potential in the APAC markets with the current, future (2030) and technical potential renewable electricity generation capacity.

The results show how much the large renewable energy projects, as the result of the respective APAC markets renewable energy policy implementation, can support the decarbonization of the industry sectors falling under the research scope.

It is then possible to answer to the main research question in the chapter 7.

Discussion

7

Today, over 70 Mt yr^{-1} of hydrogen is produced globally. Hydrogen is used as feedstock in the petrochemical industry, oil refining, and steel manufacturing. About 75% of the hydrogen dedicated production is done by the steam reforming of natural gas. 23% is produced from coal gasification while less than 2% is produced from water electrolysis. The demand for hydrogen is expected to grow, as in the future, it could be used in the transport, heat, and power sectors.

The decarbonisation of industry sectors that cannot be easily electrified can be done by replacing hydrogen produced from fossil fuels by hydrogen produced from renewable powered water electrolysis. This concept, consisting on associating a non-electrifiable industry sector with renewable energy, is described as "sector coupling".

The research work has shown that the APAC markets, with their renewable energy technical potential, could fulfil their demand for renewable hydrogen. That way, it would be possible to decarbonize the industry sectors falling under the research scope.

Except for China for which the policy instruments are being reviewed, the APAC markets are still providing substantial subsidies to promote the development of their renewable energy capacity. As a comparison, the European market has recently seen a change in the policies with the drastic reduction or the discontinuity of renewable electricity production subsidies. This has led renewable energy project developers to consider other ways of increasing the quality of their revenue streams. In this context, some European renewable energy project developers have integrated power-to-gas systems to their portfolio.

Electronic supplementary material The online version of this chapter (https://doi.org/10.1007/978-3-658-32642-5_7) contains supplementary material, which is available to authorized users.

M. Jaunatre, *Renewable Hydrogen*, Business Analytics, https://doi.org/10.1007/978-3-658-32642-5_7

Hydrogen has been identified by the APAC markets as a key element of the future global energy landscape. Policies to support the development of the hydrogen economy are developed. Australia, South-Korea, and Japan have established ambitious hydrogen strategies and roadmaps supporting the development of a hydrogen economy in the two following decades. China and Taiwan have recognized the importance of the hydrogen in the future energy landscape.

The APAC markets have different positioning in the hydrogen value chain. While Australia ambitions to be a global leader in hydrogen exportation, Japan and South-Korea are investing in the development of downstream technologies such as FCEVs, fuel cells, and other heat and power applications. China is present at all steps of the hydrogen value chain while Taiwan is not yet clearly positioned.

The research work has investigated the future potential for decarbonisation of each of the APAC markets. With the renewable electricity generation targets at 2030, the results show even if the entirety of the renewable electricity produced would be devoted to renewable hydrogen production, none of the APAC markets could cover their national hydrogen demand, at the exception of Taiwan. Consequently, if only the curtailed renewable electricity generation would be used, that would not be enough to satisfy the hydrogen demand of the APAC markets. The use of the power-to-gas systems would not be viable from an economic standpoint.

The first finding is that the current renewable energy policy frameworks of the APAC markets are not designed to support a deep decarbonisation of the industry sectors falling under the research scope. Even at the horizon 2030, the renewable electricity generation would not suffice the hydrogen demand.

Taking the analogy of the "sector coupling" described above, if the APAC markets renewable energy policy frameworks were one sector and their respective hydrogen policy frameworks were another sector, the research work shows that those two sectors are "not-coupled" or "de-coupled".

Overview
This answer the main research question: What is the influence and impact of the renewable energy policies on the decarbonisation of industry sectors using fossil fuel-based hydrogen?

Currently the influence and impact are limited to inexistent.

This implies that, unless future changes of the current policies, any additional renewable energy capacity required to produce renewable hydrogen would be subsidy-free, unfavourably increasing the price gap between the fossil fuel-based and the renewable hydrogen production.

As displayed in appendix 24, when investigating the renewable energy technical potential of the APAC markets, the research work shows that all the APAC markets could cover their own renewable hydrogen demand while operating profitably power-to-gas systems. However, it is unrealistic to assume that the APAC markets would use their full technical potential of renewable energy capacity.

In this context, the second finding is that the solution for a deep decarbonisation of the industry sectors under the research scope is international rather than national. Australia has the capacity to produce enough renewable hydrogen to cover the demand of the APAC markets, while other APAC markets, such Japan and South-Korea in line with their respective hydrogen strategies, could focus on the development of downstream applications.

Should renewable hydrogen play a central role in the future energy landscape, (1) improvement, (2) clarification, and (3) harmonization of the present renewable energy and hydrogen policy frameworks are required.

(1) The improvement of the policy frameworks could follow a path like the one under development in the Netherlands. Linking renewable energy and renewable hydrogen production into integrated tenders would achieve two goals. In a subsidy-free policy framework, it would mitigate the exposure of the large renewable energy project developers from the merchant risk. Furthermore, the tenders would be won based on an objective criterion being the lowest price of renewable hydrogen, steering the renewable hydrogen production costs down.

(2) A clarification of the policy framework regarding the origin of the hydrogen is required. With ambiguous definitions and technology neutral strategies, it is not possible to differentiate the renewable hydrogen from the fossil fuel-based hydrogen. This prevents a fair competition and possibly further premium remuneration for the renewable hydrogen.

(3) As the decarbonisation of the industry sectors falling into the research scope would find a solution at the international scale, the harmonization of the technical specifications, standards, operating procedures, and legal framework is required. This would imply coordinating the national strategies and complementing the respective APAC markets strengths to achieve a common decarbonisation goal.

Conclusion

8

The APAC countries falling under the research scope have well developed renewable energy policies, targets, and instruments. Those are designed to reduce GHG emissions, meet the Paris Agreement targets, and reduce the dependency on fossil energy.

Concurrently, the APAC countries have identified hydrogen as a key element of the future global energy landscape. Australia, Japan, and South-Korea have developed national strategies supporting the development of a hydrogen economy.

The literature review and the semi-structured interviews findings have clarified the goals and implications of the APAC markets renewable energy and hydrogen policies. It has shown that the design and the implementation of the respective policies are not linked. Moreover, the APAC hydrogen policy frameworks are often technology neutral, without specifically referring to the renewable hydrogen.

Hydrogen is a key feedstock for industrial processes such as the ammonia and methanol production, the oil refining, and the steel manufacturing. Hydrogen feedstock is produced from CO_2 emitting processes such as the steam reforming of natural gas, the coal gasification, and the oil refining by-products. Currently, the renewable hydrogen production via water electrolysis is four-fold more expensive than the fossil fuel-based hydrogen production processes.

Once the quantity of hydrogen required for the production processes of ammonia, methanol, steel, and oil refining is determined, it is possible to estimate the amount of renewable electricity required for the decarbonization of those production processes with renewable hydrogen.

The estimation of the decarbonization potential in the APAC markets with the current, future (2030) and technical potential renewable electricity generation capacity indicates how much the large renewable energy projects, as the result of the respective APAC markets renewable energy policy implementation,

M. Jaunatre, *Renewable Hydrogen*, Business Analytics, https://doi.org/10.1007/978-3-658-32642-5_8

can support the decarbonization of the industry sectors falling under the research scope.

However, the current renewable energy policy frameworks of the APAC markets are not set-up to support a deep decarbonisation of the industry sectors falling under the research scope. Even at the horizon 2030, the renewable electricity generation would not suffice the hydrogen demand.

An international effort for improving, clarifying and harmonizing the present renewable energy and hydrogen policy frameworks of the APAC markets would enable the deep decarbonisation of the industry sectors under the research scope. Country specific hydrogen strategies and renewable energy technical potential are complementary. Moreover, suitable policy frameworks are required to reduce the cost gap between the fossil fuel-based hydrogen and the renewable hydrogen.

The research work focused on the decarbonisation of the production processes of ammonia, methanol, steel and oil refining. The hydrogen demand for the transportation, heat, and power sectors has been deliberately excluded from the research scope. However, according to some of the APAC markets hydrogen strategies, those sectors are expected to generate a significant demand for hydrogen in a near future. Further research could be conducted to see how this additional hydrogen demand would impact the results obtained in this paper.

References

International Energy Agency, Iea. *Renewable Energy for Industry From Green Energy to Green Materials and Fuels Cédric Philibert*. www.iea.org/t&c/ (December 2, 2019).

No title, https://bze.org.au/wp-content/uploads/10_GW_Vision_Final-Beyond-Zero-Emissi ons-NT-2019-compressed.pdf (June 6, 2020a).

No title. https://www.thinkchina.ku.dk/documents/CREO-2019-EN-Final-0316.pdf (June 9, 2020b).

No title. https://www.enecho.meti.go.jp/en/category/others/basic_plan/5th/pdf/strategic_ene rgy_plan.pdf (June 6, 2020c).

No title. https://s3.ap-southeast-2.amazonaws.com/hdp.au.prod.app.vic-engage.files/6415/ 8025/2510/VHIP_discussion_paper_FA3_WEB_booklet.pdf (May 3, 2020d).

No title. https://www.fao.org/tempref/agl/agll/docs/ferttaiwan.pdf (April 24, 2020e).

No titlehttps://prd-wret.s3-us-west-2.amazonaws.com/assets/palladium/production/atoms/ files/mcs-2019-nitro.pdf (June 7, 2020f).

No title. https://www.energy-transitions.org/sites/default/files/CHINA_2050_A_FULLY_ DEVELOPED_RICH_ZERO_CARBON_ECONOMY_ENGLISH.pdf (June 7, 2020g).

A 1.5°C COMPATIBLE CARBON BUDGET FOR WESTERN AUSTRALIA. 2019.

About the Renewable Energy Target. https://www.cleanenergyregulator.gov.au/RET/About-the-Renewable-Energy-Target (June 6, 2020).

Adams, William C. 2015. "Conducting Semi-Structured Interviews." https://www.resear chgate.net/publication/301738442_Conducting_Semi-Structured_Interviews (December 3, 2019).

Agency, International Energy. 2015. "Technology Roadmap: Hydrogen and Fuel Cells." *SpringerReference*.

Åhman, Max et al. 2018. *Hydrogen Steelmaking for a Low-Carbon Economy A Joint LU-SEI Working Paper for the HYBRIT Project*. www.sei.org (April 10, 2020).

Alterman, Eric. 2018. "Money Well Spent." *Nation* 291(13): 10.

Andersson, Joakim, and Stefan Grönkvist. 2019. "Large-Scale Storage of Hydrogen." *International Journal of Hydrogen Energy* 44(23): 11901–19.

Arabzadeh, Vahid, Sannamari Pilpola, and Peter D. Lund. 2019. "Coupling Variable Renewable Electricity Production to the Heating Sector through Curtailment and Power-to-Heat

© The Editor(s) (if applicable) and The Author(s), under exclusive license to Springer Fachmedien Wiesbaden GmbH, part of Springer Nature 2021
M. Jaunatre, *Renewable Hydrogen*, Business Analytics,
https://doi.org/10.1007/978-3-658-32642-5

Strategies for Accelerated Emission Reduction." *Future Cities and Environment* 5(1): 1–10.

Au, Climatecouncil Org. 2018. *POWERING PROGRESS: STATES RENEWABLE ENERGY RACE.*

AURES Consortium. 2019. "Effect of Auctions on Financing Conditions for Renewable Energy." (April).

Australia's Hydrogen Strategy Has Officially Landed I Energy Magazine. https://www.energy magazine.com.au/australias-hydrogen-strategy-has-officially-landed/ (June 6, 2020).

Basic Hydrogen Strategy Determined (METI). https://www.meti.go.jp/english/press/2017/1226_003.html (June 6, 2020).

Bhattacherjee, Anol. *Scholar Commons Social Science Research: Principles, Methods, and Practices.* https://scholarcommons.usf.edu/oa_textbooks/3.

Bird, Lori et al. 2016. "Wind and Solar Energy Curtailment: A Review of International Experience." *Renewable and Sustainable Energy Reviews* 65: 577–86.

Blazquez, Jorge, Rolando Fuentes-Bracamontes, Carlo Andrea Bollino, and Nora Nezamuddin. 2018. "The Renewable Energy Policy Paradox." *Renewable and Sustainable Energy Reviews* 82(April 2017): 1–5. http://dx.doi.org/https://doi.org/10.1016/j.rser.2017.09.002.

Brau, Jean-Florian. 2013. "Production of Hydrogen for Oil Refining by Thermal Gasification of Biomass: Process Design, Integration and Evaluation." *Heat and Power Technology* (May): 71.

Bureau of Energy, Ministry of Economic Affairs, R.O.C. 2009. "Four-Year Plan of Promotion for Wind Power." *Bureau of Energy, Ministry of Economic Affairs, R.O.C.* (c): 2025. https://www.moeaboe.gov.tw/ECW/english/content/SubMenu.aspx?menu_id=1525.

Burrell, Gibson, Gareth Morgan, Gibson Burrell, and Gareth Morgan. 2019. "Assumptions about the Nature of Social Science." *Sociological Paradigms and Organisational Analysis*: 1–9.

Carmo, Marcelo, and Detlef Stolten. 2018. "Energy Storage Using Hydrogen Produced from Excess Renewable Electricity: Power to Hydrogen." In *Science and Engineering of Hydrogen-Based Energy Technologies: Hydrogen Production and Practical Applications in Energy Generation*, Elsevier, 165–99.

China. *Energy Transition Trends 2019.*

CNREC. 2019. "China Energy Policy Newsletter." https://www.energypartnership.cn/filead min/user_upload/china/media_elements/newsletter/CNREC_Newsletter_EN/China_Ene rgy_Policy_Newsletter_March_2019.pdf (June 7, 2020).

Collodi, Guido, Giuliana Azzaro, Noemi Ferrari, and Stanley Santos. 2017. "Techno-Economic Evaluation of Deploying CCS in SMR Based Merchant H2 Production with NG as Feedstock and Fuel." In *Energy Procedia*, Elsevier Ltd, 2690–2712.

Commonwealth of Australia. 2019. *AUSTRALIA'S NATIONAL HYDROGEN STRATEGY.*

Couture, Toby D, David Jacobs, Wilson Rickerson, and Victoria Healey. 2015. *The Next Generation of Renewable Electricity Policy: How Rapid Change Is Breaking Down Conventional Policy Categories.* www.nrel.gov/publications. (June 6, 2020).

Dalla Riva, Alberto. 2016. *System Value of Wind Power-An Analysis of the Effects of Wind Turbine Design Economic Dispatch Modelling of Medium-Term System Implications of Advanced Wind Power Technologies.* www.ienie.dii.unipd.it (June 7, 2020).

Davis, Steven J et al. "Net-Zero Emissions Energy Systems."

Deng, Xue, Hewu Wang, Haiyan Huang, and Minggao Ouyang. 2010. "Hydrogen Flow Chart in China." *International Journal of Hydrogen Energy* 35(13): 6475–81. http://dx.doi.org/ https://doi.org/10.1016/j.ijhydene.2010.03.051.

Department of Premier and Cabinet. https://www.dpac.tas.gov.au/divisions/climatechange/ tasmanias_climate_change_action_plan_20172021/advancing_our_renewable_energy_ capability (June 6, 2020).

Department of Primary Industries and Regional Development. 2019. *Western Australian Renewable Hydrogen Strategy H 2 H 2.*

Eastspring. "Hydrogen- Powering South Koreas Future - Darren Choi." https://www.eastsp ring.com/insights/hydrogen-powering-south-koreas-future (December 18, 2019).

Eisenhardt, Kathleen M. 1989. 14 Source: The Academy of Management Review *Building Theories from Case Study Research.*

Enel. 2016. "Analysis of Key Factors for Successful Auction Programs: Experiences Outside of Europe."

ENERGY STATISTICS 能源統計 手冊 中華民國107年 經 濟 部 能 源 局 *BUREAU OF ENERGY, MOEA.* 2019.

EPA. 2008. "Technical Support Document for Co 2 Supply : Proposed Rule for Mandatory Reporting Of." : 1–30.

Erisman, Jan Willem et al. 2008. "How a Century of Ammonia Synthesis Changed the World." *Nature Geoscience* 1(10): 636–39.

Fargere, Alena et al. 2018. "Hydrogen an Enabler of the Grand Transition. Future Energy Leader Position Paper 2018." www.worldenergy.org/wec-.

Fasihi, Mahdi, and Christian Breyer. 2019. "Baseload Electricity and Hydrogen Supply Based on Hybrid PV-Wind Power Plants." *Journal of Cleaner Production* 243: 118466. https:// doi.org/https://doi.org/10.1016/j.jclepro.2019.118466.

Federal Register of Legislation - Australian Government. https://www.legislation.gov.au/Ser ies/C2004A00767 (June 6, 2020).

Formulation of a New Strategic Roadmap for Hydrogen and Fuel Cells. https://www.meti.go. jp/english/press/2019/0312_002.html (June 6, 2020).

From 0 to 15GW by 2030: Four Reasons Why Taiwan Is the Offshore Wind Market in Asia | Global Wind Energy Council. https://gwec.net/from-0-to-15gw-by-2030-four-reasons-why-taiwan-is-the-offshore-wind-market-in-asia/ (June 6, 2020).

GENI. 2012. *How Is 100% Renewable Energy Possible in Japan by 2020?* www.geni.orgpet er@geni.org (June 7, 2020).

Van Gerwen, Rob, Marcel Eijgelaar, and Theo Bosma. 2019. "Hydrogen in the Electricity Value Chain."

Glenk, Gunther, and Stefan Reichelstein. 2019. "Economics of Converting Renewable Power to Hydrogen." *Nature Energy* 4(3): 216–22.

Global Solar Atlas. https://globalsolaratlas.info/map?c=11.523088,8.173828,3 (June 7, 2020).

Global Wind Atlas. https://globalwindatlas.info/ (June 7, 2020).

Government, A C T. 2020. "Act Sustainable."

Government, NSW. *NSW Electricity Strategy Our Plan for a Reliable, Affordable and Sustainable Electricity System.*

Government of Korea. 2019. "Hydrogen Economy Roadmap of Korea." (January).

Government of South Australia. 2019. "South Australia's Hydrogen Action Plan."

Green Hydrogen Opportunities in Selected Industrial Processes Workshop Summary Report. https://ec.europa.eu/jrc (April 5, 2020).

Guest, Greg, Arwen Bunce, and Laura Johnson. 2006. "How Many Interviews Are Enough?" *Field Methods* 18(1): 59–82. http://journals.sagepub.com/doi/https://doi.org/10.1177/152 5822X05279903 (April 18, 2020).

Hagaman, Ashley K., and Amber Wutich. 2017. "How Many Interviews Are Enough to Identify Metathemes in Multisited and Cross-Cultural Research? Another Perspective on Guest, Bunce, and Johnson's (2006) Landmark Study." *Field Methods* 29(1): 23–41. http://journals.sagepub.com/doi/10.1177/1525822X16640447 (April 18, 2020).

Heiligtag, Sven, Florian Kühn, Florian Küster, and Joscha Schabram. 2018. *Merchant Risk Management: The New Frontier in Renewables.*

Hong, Jong Ho et al. 2019. "Long-Term Energy Strategy Scenarios for South Korea: Transition to a Sustainable Energy System." *Energy Policy* 127: 425–37.

Hydrogen Energy Supply Chain | Making a Global Hydrogen Supply Chain between Australia and Japan a Reality. https://hydrogenenergysupplychain.com/ (June 6, 2020).

Hydrogenics. 2018. *Introduction Video.* https://youtu.be/UJXhX4dLMtA (April 6, 2020).

IEA. 2018a. "Australia – 2018 Update Bioenergy Policies and Status of Implementation." https://www.ieabioenergy.com/wp-content/uploads/2018/10/CountryReport2018_Australia_final.pdf (June 7, 2020).

IEA. 2018b. "Japan—2018 Update Bioenergy Policies and Status of Implementation." https://www.ieabioenergy.com/wp-content/uploads/2018/10/CountryReport2018_Japan_final.pdf (June 7, 2020).

IEA. 2018c. "Korea—2018 Update Bioenergy Policies and Status of Implementation." https://www.ieabioenergy.com/wp-content/uploads/2018/10/CountryReport2018_Korea_final.pdf (June 7, 2020).

International ASA, Yara. 2018. *Fertilizer Industry Handbook 2018.*

International Renewable Energy Agency. 2012. *RENEWABLE ENERGY TECHNOLOGIES: COST ANALYSIS SERIES Volume 1: Power Sector Acknowledgement.* www.irena.org/Publications (June 7, 2020).

International Renewable Energy Agency. 2014. *Renewable Energy Prospects: China.* www.irena.org/remap (April 24, 2020).

International Renewable Energy Agency. 2019. "Renewable Energy Auctions: Status and Trends Beyond Price." : 32.

Jacobs. 2019. *Hydrogen Economy Australia's Pursuit of a Large Scale Evaluating the Economic Viability of a Sustainable Hydrogen Supply Chain Model.*

Jacobs, Jeffrey. 2016. "Economic Modeling of Cost Effective Hydrogen Production From Water Electrolysis by Utilizing Iceland's Regulating Power Market Economic Modeling of Cost Effective Hydrogen Production From Water Electrolysis by Utilizing Iceland's Regulating Power Mark." (January): 1–46. https://skemman.is/bitstream/1946/23812/1/Jeffrey Jacobs.pdf.

Jaunatre, Matthieu. 2020a. "Introduction Video for Renewable Energy and Renewable Hydrogen Experts in the Asia Pacific Region—YouTube." https://www.youtube.com/watch?v=gEKESyrpHvM&feature=youtu.be (June 15, 2020).

Jaunatre, Matthieu. 2020b. "Introduction Video for Renewable Energy Project Developers in the Asia Pacific Region—YouTube." https://www.youtube.com/watch?v=Sw3sOrvx2hI (June 15, 2020).

Jaunatre, Matthieu. 2020c. "Introduction Video for Renewable Hydrogen Offtakers in the Asia Pacific Region—YouTube." https://www.youtube.com/watch?v=o3hZL5rIfF8 (June 15, 2020).

Joos, Michael, and Iain Staffell. 2018. "Short-Term Integration Costs of Variable Renewable Energy: Wind Curtailment and Balancing in Britain and Germany." *Renewable and Sustainable Energy Reviews* 86(January): 45–65.

Kær, Søren Knudsen; Al Shakhshir, Saher. 2016. "Power2Hydrogen WP1 Potential of Hydrogen in Energy Systems." : 1–66. https://hybalance.eu/wp-content/uploads/2017/01/Power2Hydrogen-WP1-report-Potential-of-hydrogen-in-energy-systems.pdf.

KEA - KOREA ENERGY AGENCY. https://www.energy.or.kr/renew_eng/new/renewable.aspx (June 6, 2020).

Kwasi-Effah, Collins et al. 2015. "A Review on Electrolytic Method of Hydrogen Production from Water." *American Journal of Renewable and Sustainable Energy* 1(2): 51–57. https://www.aiscience.org/journal/ajrse.

Lacerda, Juliana Subtil, and Jeroen C.J.M. van den Bergh. 2016. "Mismatch of Wind Power Capacity and Generation: Causing Factors, GHG Emissions and Potential Policy Responses." *Journal of Cleaner Production* 128: 178–89.

Lambert, Susan. 2008. "For Business Model Research." (June 2008): 277–89.

Langworthy, Alan et al. 2017. *Roadmap to Renewables Fifty per Cent by 2030 Northern Territory Chair-Expert Panel.*

Sanghoon, Lee. 2019. *Renewable Energy 3020 Plan and Beyond.*

van Leeuwen, Charlotte, and Machiel Mulder. 2018. "Power-to-Gas in Electricity Markets Dominated by Renewables." *Applied Energy* 232(October): 258–72. https://doi.org/10.1016/j.apenergy.2018.09.217.

Levi, Peter and Jonathan M Cullen. 2018. "Mapping Global Flows of Chemicals: From Fossil Fuel Feedstocks to Chemical Products." https://pubs.acs.org/sharingguidelines (January 27, 2020).

Lewandowska-Bernat, Anna, and Umberto Desideri. 2017. "Opportunities of Power-to-Gas Technology." *Energy Procedia* 105: 4569–74. https://doi.org/10.1016/j.egypro.2017.03.982.

Barriball, Louise and While, Alison. 1994. 19 Journal of Advanced Nursing *Collecting Data Using a Semi-Structured Interview: A Discussion Paper.*

Shyi-Min, Lu. 2016. "A Review of Renewable Energies in Taiwan IJESRT INTERNATIONAL JOURNAL OF ENGINEERING SCIENCES & RESEARCH TECHNOLOGY A REVIEW OF RENEWABLE ENERGIES IN TAIWAN." © *International Journal of Engineering Sciences & Research Technology.* https://www.ijesrt.com.

Luo, Xing, Jihong Wang, Mark Dooner, and Jonathan Clarke. 2015. "Overview of Current Development in Electrical Energy Storage Technologies and the Application Potential in Power System Operation." *Applied Energy* 137: 511–36. https://doi.org/10.1016/j.apenergy.2014.09.081.

Matijašević, Lj, and M Petrić. 2016. "Integration of Hydrogen Systems in Petroleum Refinery." *Chem. Biochem. Eng. Q* 30(3): 291–304.

McKinsey & Company. 2018. "Hydrogen Roadmap Korea." (November).

MMSA. 2012. *Chapter V—Regional Methanol Market Analysis—World REGIONAL.*

Dr. Elangovan, Dr. Rajendran. 2015. "Conceptual Model : A Framework for Institutionalizing the Vigor in Business Research." (April 2015).

Ohira, Eiji. 2019. "Japan Policy and Activity on Hydrogen Energy." 926.

OPPORTUNITIES FOR AUSTRALIA FROM HYDROGEN EXPORTS ACIL ALLEN CON-SULTING FOR ARENA. 2018.

"Orsted Participates in Tender for Holland Coast South 3–4 Offshore Wind Farm." https://orsted.com/en/media/newsroom/news/2019/03/orsted-participates-in-ten der-for-holland-coast-south-3-4-offshore-wind-farm (June 6, 2020).

Otto, Alexander et al. 2017. "Power-to-Steel: Reducing CO_2 through the Integration of Renewable Energy and Hydrogen into the German Steel Industry." *Energies* 10(4).

De Pee, Arnout et al. 2018. *(No Title).*

Pilbara Green Hydrogen Project Grows to 15GW Wind and Solar | RenewEconomy. https://reneweconomy.com.au/pilbara-green-hydrogen-project-grows-to-15gw-wind-and-solar-97972/ (June 6, 2020).

Presentation: Renewable Energy for Business Simplified Nov 2019 - Australian Renewable Energy Agency (ARENA). https://arena.gov.au/knowledge-bank/presentation-renewable-energy-for-business-simplified-nov-2019/ (June 6, 2020).

Queensland Government. 2019. "Queensland Hydrogen Industry Strategy." (May). www.dsd mip.qld.gov.au.

Rabiei, Zahra. 2012. "HYDROGEN MANAGEMENT IN REFINERIES." *Petroleum & Coal* 54(4): 357–68. www.vurup.sk/petroleum-coal (April 5, 2020).

Ramsden, Todd et al. 2013. "Hydrogen Pathways: Updated Cost, Well-to-Wheels Energy Use, and Emissions for the Current Technology Status of Ten Hydrogen Production, Delivery, and Distribution Scenarios, NREL/TP-6A10–60528." (March): 1–268. www.nrel.gov/pub lications.

Renewable Capacity Statistics 2019. https://www.irena.org/publications/2019/Mar/Renewa ble-Capacity-Statistics-2019 (April 19, 2020).

The State of Victoria Department of Environment, Land, Water and Planning. 2017. *Renewable Energy Action Plan.*

Renewable, International, and Energy Agency. *Zhangjiakou Energy Transformation Strategy 2050: Pathway to a Low-Carbon Future.*

Republic of Korea | Iphe. https://www.iphe.net/republic-of-korea (June 6, 2020).

Rosetti, Valentina. 2007. "Catalysts for H2 Production." : 1–294. https://amsdottorato.unibo. it/427/.

Salt, Stuart. 2018. "Renewable Energy In Taiwan." https://www.conventuslaw.com/report/ren ewable-energy-in-taiwan/ (June 7, 2020).

Saygın, Deger, Martın K Patel, Cecılıa Tam, and Dolf J Gıelen. *ChemiCal and PetroChemiCal SeCtor Potential of Best Practice Technology and Other Measures for Improving Energy Efficiency IEA InformAtIon PApEr.* www.iea.org/about/copyright.asp (April 4, 2020).

Schmidt et al. 2017. "Future Cost and Performance of Water Electrolysis: An Expert Elici-tation Study." *International Journal of Hydrogen Energy* 42(52): 30470–92. https://doi. org/https://doi.org/10.1016/j.ijhydene.2017.10.045.

Shiva Kumar, Sampangi and Himabindu. 2019. "Hydrogen Production by PEM Water Electrolysis – A Review." *Materials Science for Energy Technologies.*

Simons, Andrew, and Christian Bauer. 2011. Transition to Hydrogen: Pathways Toward Clean Transportation *Life Cycle Assessment of Hydrogen Production.*

Staffell, Iain et al. 2019. "The Role of Hydrogen and Fuel Cells in the Global Energy System." *Energy and Environmental Science* 12(2): 463–91.

State of Tasmania. 2019. "Tasmanian Renewable Hydrogen Action Plan." https://www.stateg
  rowth.tas.gov.au/__data/assets/pdf_file/0003/207705/Draft_Tasmanian_Hydrogen_Act
  ion_Plan_-_November_2019.pdf (June 6, 2020).
STEEL STATISTICAL YEARBOOK 2019 Concise Version.
Taylor, Peter Charles, Elisabeth Lily Taylor, and Bal Chandra Luitel. 2012. "Multi-
  Paradigmatic Transformative Research as/for Teacher Education: An Integral Perspec-
  tive." In Second International Handbook of Science Education, Springer Netherlands,
  373–87.
Thanapalan, Kari et al. 2012. "Development of Energy Saving Mechanism for Renewable
  Hydrogen Vehicles." Renewable Energy and Power Quality Journal 1(10): 480–85.
The Future of Hydrogen. 2019. The Future of Hydrogen (June).
The Victorian Hydrogen Investment Program :: Engage Victoria. https://engage.vic.gov.au/
  vhip (June 6, 2020).
TIQ Australia. 2019. TIQ International Market Report Opportunities for Queensland
  Businesses in Japan's Hydrogen Market.
Torabi, Roham, Alvaro Gomes, and F. Morgado-Dias. 2018. "The Duck Curve Characte-
  ristic and Storage Requirements for Greening the Island of Porto Santo." Energy and
  Sustainability in Small Developing Economies, ES2DE 2018 - Proceedings (October):
  13–20.
Townsend, Keith. Saturation And Run Off: How Many Interviews Are Required In Qualitative
  Research?
Trends of Renewable Energy Capacity in Japan Institute for Sustainable Energy Policies.
UNSD—Energy Statistics. https://unstats.un.org/unsd/energystats/pubs/balance/ (March 8,
  2020).
UTS. 2016. "100% RENEWABLE ENERGY FOR AUSTRALIA Decarbonising Austra-
  lia's Energy Sector Within One Generation Prepared for: GetUp! And Solar Citi-
  zens." https://www.uts.edu.au/sites/default/files/article/downloads/ISF_100%25_Austra
  lian_Renewable_Energy_Report.pdf (June 7, 2020).
Verheul, Bente. 2019. "Overview of Hydrogen and Fuel Cell Developments in China."
  Holland Innovation Network in China (January). https://www.nederlandwereldwijd.nl/
  binaries/nederlandwereldwijd/documenten/publicaties/2019/03/01/waterstof-in-china/
  Holland+Innovation+Network+in+China+-+Hydrogen+developments.+January+2019.
  pdf.
VICTORIAN RENEWABLE ENERGY TRANSITION ECONOMIC IMPACTS MODELLING.
  2019.
Vogl, Valentin, Max Åhman, and Lars J. Nilsson. 2018. "Assessment of Hydrogen Direct
  Reduction for Fossil-Free Steelmaking." Journal of Cleaner Production 203: 736–45.
Wang, Lu et al. 2018. "Greening Ammonia toward the Solar Ammonia Refinery." Joule 2(6):
  1055–74. https://doi.org/10.1016/j.joule.2018.04.017.
Wang, Ren-chain. 2017. "Current Thrusts to 2025 Renewable Energy Targets in Taiwan."
Western Australia to Consume Its Carbon Budget 20 Years Too Early, Report Finds—Pv
  Magazine Australia. https://www.pv-magazine-australia.com/2019/11/27/western-austra
  lia-to-consume-its-carbon-budget-20-years-too-early-report-finds/ (June 9, 2020).
What's New - News & Releases—Ministry of Economic Affairs, R.O.C. https://www.moea.
  gov.tw/MNS/english/news/News.aspx?kind=6&menu_id=176&news_id=83360 (June 6,
  2020).

World Bank Group. 2020. "Technical Potential for Offshore Wind in China." https://doc
    uments.worldbank.org/curated/en/749651586838852952/pdf/Technical-Potential-for-
    Offshore-Wind-in-China-Map.pdf (June 7, 2020).
Xiaoying, Xu et al. 2017. "Clean Coal Technologies in China Based on Methanol Platform."
    Catalysis Today 298(December): 61–68. https://doi.org/10.1016/j.cattod.2017.05.070.
Yang, Jianbo, Qunyi Liu, Xin Li, and Xiandan Cui. 2017. "Overview of Wind Power in China:
    Status and Future." *Sustainability (Switzerland)* 9(8): 1–12.

The manufacturer's authorised representative in the EU is Springer
Nature Customer Service Centre GmbH, Europaplatz 3, 69115 Heidelberg,
Germany. If you have any concerns regarding our products, please
contact ProductSafety@springernature.com

Printed and bound by CPI Group (UK) Ltd, Croydon, CR0 4YY
28/04/2026
02098481-0006